Human Ability
Genetic and Environmental Influences

Proceedings of a Conference organised by the Galton Institute, London, 2005

Edited by

Robert Peel and Mazin Zeki

PUBLISHED BY THE GALTON INSTITUTE

British Library Cataloguing in Publication Data

Human ability : genetic and environmental influences : proceedings of a conference organised by the Galton Institute, London, 2005

1.Intellect - Genetic aspects - Congresses
2.Intellect - Social aspects - Congresses

I.Peel, Robert II.Zeki, Mazin III.Galton Institute (London, England)

155.5'7

ISBN-10: 0954657004

First published 2006 by the Galton Institute, 19 Northfields Prospect, Northfields, London SW18 1PE

Printed and bound in Great Britain by CM Print, 61 Station Road, Portslade, Brighton, BN41 1DF.

Contents

Notes on the Contributors

Professor Ian Deary, University of Edinburgh

Professor Douglas K Detterman, Department of Psychology, Case Western Reserve University, Cleveland, OH 44106. E-mail: Detterman@case.edu.

Professor Robin Dunbar, School of Biological Sciences, University of Liverpool, Liverpool L69 7ZB. E-mail: rimd@liv.ac.uk.

Professor Joan Freeman, Middlesex University, Department of Lifelong Learning and Education, London, UK

Professor Louise Higgins, University College, Chester

Preface

This book contains the proceedings of the forty-first annual conference of the Galton Institute, the most recent in a series which was begun in 1964 in fulfilment of a decision by Council that the main activity of the Society [as the Institute then was] should be devoted to the common ground between the biological and social sciences and bring together for the mutual exchange of ideas and information those interested in the relative importance of genetic and environmental influences.

In 1964-65-66 a trio of successful symposia were organised respectively on "Biological Aspects of Social Problems", "Genetic and Environmental Factors in Human Ability" and "Social and Genetic Influences on Life and Death". In a review of the 1965 symposium proceedings, Professor A. H. Halsey commended the biosocial aspirations of the Institute:

> "A hundred years ago" he suggested "this would have given rise to no excitement. Spencerian unity would have been taken for granted. But, as is well known, these two partners of nineteenth century progress parted company in the early years of the twentieth century. Separation was ideological and political rather than intellectual. Most sociologists today have been taught to assume, mainly through second-hand references to the excesses of social Darwinism, that application of biological ideas to human society is the instrument of scoundrels designed to justify predatory ethics and to defend racial, ethnic, caste and class inequalities."

This divorce, Halsey believed, had been costly and had prevented the emergence of a highly appropriate modern form of liberal education which, if it existed, would reveal the over simplification of the so-called 'two cultures' problem. Similarly, in a review of the subsequent 1966 symposium volume Jean Sutter suggested that the Institute had "thus revitalised the study of human problems inherent in our kind of civilisation, which, if they are not resolved, are capable of menacing evolution itself".

It is purely accidental that the title of the 2005 conference corresponds so closely with that of 1965; there are few members of the present Council who can have remembered that earlier meeting. Rather, the coincidence of the two titles reflects the consistency with which the Institute has adhered to the biosocial brief laid down forty years ago. A comparison of the content of the two volumes might be expected to reveal the extent to which the rapprochement anticipated by Council has proceeded over a period of four decades.

The participants in the 1965 book reflected the hegemony of the environmentalists in the nature/nurture debate. Sociologists were well represented – by Bernstein and Young, Jean Floud and others – who clearly reflected the "blank slate" theory of education as well as their desire to employ the schools as instruments of social engineering and Lord Robbins, who has much to answer for in the history of British education, argued – somewhat incoherently – that if tests showed that fewer numbers than his committee had proposed were likely to benefit from higher education then those tests must be made "much more reliable".

By 2005 sociology was unrepresented and psychology which, with demography, is the "hardest" of the social sciences employing the natural science techniques of experiment and measurement to social behaviour and social processes, was to the forefront. The role of heredity in human ability was taken for granted and its contribution of 50 to 70 per cent more or less generally accepted; after all the intervening years had seen the rise of sociobiology, the costly failure of affirmative action programmes, especially in the United States, the restoration of the reputation of Sir Cyril Burt following his posthumous ordeal by fire and the findings of the Minnesota Twin Studies.

Papers in this volume describe important contemporary research in the field of human ability and the Institute is grateful to the distinguished speakers who contributed to the conference and the book. Our thanks, too, to Mrs Betty Nixon for her help with both.

Conference Organiser's Introduction

Mazin Zeki

Since Francis Galton wrote *Hereditary Genius*, many have followed in his footsteps. Galton himself later admitted that Genius was perhaps a poorly chosen term.

But in the hundred years since Spearman's seminal essay on *g*, the London School, which has included many eminent Institute members, has gone from strength to strength. Although it has not totally prevailed, it has decisively shifted the debate to which this volume is a modest contribution.

It seems almost too obvious to mention, that intelligence has been the motor of human advance. Intelligence is also becoming the motor of human division and the hidden tool of competition, although carefully camouflaged or not commented on. This extended to public disputes within the academic community, which has insisted on often neutral and evasive conclusions.

As Douglas Detterman notes, we may be becoming more segregated by intelligence. This is not only because intelligence is a marker for income and social status (with some important exceptions) – links between intelligence and other aspects of human ability have been ignored for too long. Because of the prominent role of inheritance in determining intelligence, the segregation has been self-reinforcing, both between social groups and within them. This has extended to the role of intelligence in mate choice, which has been one of the dominant themes of evolutionary psychology.

Studies among humans and primates have highlighted the consistent role of the outward correlates of intelligence, such as wealth and power, as determinants in mate selection. Such patterns of mate selection increase the likelihood of inherited differences in intelligence becoming polarised, subject perhaps to mitigation by reversion to the mean as described by Francis Galton.

The papers in this volume also illustrate the changing nature of the debate. Although the traditional Chinese civil service examinations were abolished a hundred years ago, the rich heritage of the Chinese interest in intelligence was still operating – as Louise Higgins underlines. The Chinese experience is important because it is the longest continuous higher civilisation in existence with extant historical records.

When a Chinese student came to study with Spearman, he brought together for the first time the new school founded by Wundt (with which Spearman was linked), the London School and the Chinese heritage of the traditional civil service examinations. These examinations had a wide influence and were copied by adjacent societies in the Far East like Japan and Korea, as well as forming the basis for the modern examination system in Britain.

Controversy over the assessment of cognitive ability seems to have lessened considerably with the passage of time. Significantly, when Arthur Jensen toured India and China, thirty years after his Harvard Education Review article, he encountered no hostility but received an enthusiastic response.

Tests and examination systems need and demand success for some, which must mean failure for others. But exams are not the only way of assessing intelligence, although perhaps the most important and popular. Education itself has been in crisis because intelligence has been ignored on the basis of 'all must have prizes'. The competitive principle was being undermined even while it was being promoted. Far from helping the intended targets of kindness, it has fatally betrayed them. Discussions of intelligence have confused equality before the law and equal opportunity with attempts to create a pretended equality of outcome while leaving economic realities undisturbed.

Perhaps the only possibility of resolving disputes is restoring the primacy of scientific evidence through longitudinal studies, such as the one by Ian Deary demonstrating the resilience of IQ over a lifetime, or through advances in cognitive neuroscience that demonstrate the biological basis of differences in intelligence.

Giftedness in the Long Term[1]

Professor Joan Freeman

Abstract

This more than three-decade study is an in-depth comparison of labelled gifted, unlabelled but equally able gifted and random children, each triad in the same school class. It has shown how and why some developed their abilities into satisfying adult achievement but others did not. While at school, extremely high grades could be attained by the academically gifted who enjoyed teacher and parent support. But extreme precocity, intelligence scores which hit the ceiling of the tests, top marks and grade-skipping were not a passport to recognisable giftedness in the adult world. Reactions to circumstances were always individual and not necessarily related to giftedness: negative influences on development could overwhelm some individuals whereas others flew above the challenges. Childhood emotional problems were often attributed to giftedness and were found more frequently among the labelled gifted at a high level of significance. These often vanished with increasing independence, though for some they remained to darken their adulthood. The effective dynamics for success at all levels of ability were keenness, hard work, educational opportunity, emotional support and a positive emotional outlook – and chance.

Giftedness in the Long Term

The major benefit of longitudinal studies of gifted and talented children is tracking behaviour as it develops so that early indicators may be recognised and successful developmental procedures promoted for the benefit of others. The major debit is that such studies inevitably started a long time ago when things were

[1] The Freeman Follow-up Study is generously supported by the Esmée Fairbairn Foundation, UK.

different, bringing into question the relevance of findings to current circumstances. In research terms, older methodology is always old-fashioned, in the sense of - I wouldn't start from there if I were you.

Giftedness is a social construct, and this can be seen in the selection of samples of children seen as gifted. Virtually all follow-up studies of gifted children select those chosen by extremely high scores on IQ or other attainment tests, that is children who are demonstrating recognisable giftedness acceptable within a society at that time. This limits the generalisability of predictions from such samples. Subotnik, Kassan, Summers & Wasser (1993) have shown that giftedness may take many different forms; it may appear in quite unexpected situations and at different points during a lifetime. It is not always possible to identify future gifts, which means that theories and educational programmes designed for children who are precocious in conventional areas may well miss those whose gifts do not fit either now or in the future.

Attempting to avoid the trap of selection by achievement, the Fullerton Longitudinal study in California began with 130 one-year-olds of unknown potential and their families; the only criterion being that they were healthy (Gottfried, Gottfried & Guerin, in press). Measures of intellectual, physical and social development were taken regularly from 1979 to 1997. Those with an IQ of 130 or more on the Wechsler Intelligence Test were eventually deemed gifted and compared with the others. Early indications of giftedness were discovered and parents proved to be good judges. The researchers concluded that giftedness is a developmental phenomenon, which can rise – and fall – over time so that 'late bloomers' can be missed in a single testing.

Population statistics do not provide entirely satisfactory controls for longitudinal studies in gifted development because they are not focussed on the subject matter (Freeman, 1998). Yet longitudinal studies of gifted children almost never make any comparisons with control groups undergoing the same experiences and matched for age, sex, educational experience and socio-economic-level. This was true, for example, in the Terman Californian studies, which in

1925 selected 856 boy and 672 girl "geniuses" of IQ 130+, eventually producing more than 4000 variables (Terman, 1925-1929). Even for those days there were considerable flaws in the sampling, e.g. "no private, parochial (religious) or Chinese schools" (Holahan & Sears, 1995, P. 11). The subjects, aged between 2 and 22, were almost entirely the progeny of white university staff, sometimes of Terman's colleagues, collected over a period of 7 years. As early as 1928 a quarter of the original sample had been replaced, and this replacement continued for many years, so that the sample was not in fact either longitudinal or valid. But it was, of course, interesting.

Terman's "geniuses" were considerably above-average in every way, including height and leadership qualities, probably because they enjoyed well above the population norms of nourishment, exercise and education. Holahan & Sears found that the 'Termites' in their seventies and eighties were no more successful in adulthood than if they had been randomly selected from the same socio-economic backgrounds – regardless of their IQ scores. This was somewhat mirrored in the findings of Subotnik, Kassan, Summers & Wasser (1993) who investigated a sample of 210 New York children selected for the Hunter College Elementary School by their high-IQ scores (mean IQ 157). None had reached eminence by the ages of 40 to 50, nor were they any more successful than their socio-economic and IQ peers in spite of their tailor-made gifted education.

The Seattle Longitudinal Study has been concerned with intelligence and aging since 1956 (Schaie, 2005). It has examined expanding families over three generations (the constantly replenished sample reached 6000) and found that social effects influence the stability of IQ with increasing age. Intellectual and perceptual abilities remain high for individuals who stay active and open-minded; notably, people satisfied with their accomplishments in mid-life are at a considerable advantage as they age. The Munich Longitudinal Study of Giftedness began in 1985 with a sample of 26,000 children, identified on a wide variety of intellectual, personality and achievement tests (Perleth & Heller, 1994). The team devised 30 identification scales, which disclosed a significant number of gifted under-achievers, who were typically found to be more anxious, easily distracted and with lower self-esteem than the high achievers.

In a review of 14 American and German follow-up studies of varied design, Arnold and Subotnik (1994) pointed to several important factors in conditions for the development of talent. Timing, they suggested, is the "inextricable link" in the identification of potential because of age-related stages of development. Thus, the older the sample the more reliable the prediction. But for the greatest reliability information should be collected at different points in an individual's life, at best within specific subject areas in which the child shows promise and interest.

Further evidence that high level school achievement may not follow-on in adult life comes from a 15-year follow-up of 82 'valedictorians' (the highest grade earners in high school) from 32 schools across Illinois (Arnold, 1995). It showed that even such exceptional grades did not make good long-term predictors of later high achievement. Each individual was given five or six interviews after leaving school. They had enjoyed all aspects of school and had used it efficiently to prepare for their future lives. Their major academic advantage was in their determination to better themselves. Neither boys nor girls felt themselves to be outstandingly clever nor had they been labelled as such. None of this sample made outstanding progress in their careers (particularly the women), and by 26 years-old many were disillusioned.

Nor are the long-term benefits of early special provision for the gifted certain. In spite of an initially higher measured achievement and student feelings of satisfaction, the advantage of gifted education tends to disappear over a few years (White, 1992). Without the long-term perspective, programmes for the gifted may not be justified (Freeman, 2002). For example, a recent UK review of international research on Accelerated Learning found evidence of its effectiveness to be scientifically poor (Comford Boyes, Reid, Brain & Wilson, 2004). Additionally, the programme was found to be "voraciously marketed" and a placebo effect was detected. Yet for many schools around the world it is the program of choice for the gifted and talented. The question to be asked of all such programmes is how much of the initial boost to achievement is due to the Hawthorn effect, that is to sheer attention and change, and whether the effects last over years.

Measures of giftedness in the long term

The measurement of intelligence is among the best and most resilient success stories in all scientific psychology, according to the American Psychological Association's task force (Neisser, Boodoo, Bouchard, Boykin, Brody, Ceci, S.J., 1996). After a century of solid, replicated research, intelligence levels, the report concluded, reliably predict life outcomes in education and the workplace as well as aspects of health, such as how long people live. For example, a step up of just one standard deviation in IQ in 11 year-old girls improves their chances of reaching the age of 76 by 25% (Whalley & Deary, 2001).

On Wednesday, June 1, 1932, practically every Scottish child born in 1921 (N = 89,498) took the same intelligence test (the Moray House) with the same time limit after hearing the same instructions (Deary, Whiteman, Starr, Whalley & Fox, 2004). The still ongoing study of their lives is concerned with the stability of intelligence differences across the life span, the determinants of cognitive change from childhood to old age and the impact of childhood intelligence on health and quality of life in old age. Data were compared with public records for the whole UK. In this case, as a whole population was sampled it does make more sense to use the national statistics for comparison. IQ has been found strongly stable across the lifespan. Current tests and interviews show those of higher intelligence to be both physically and mentally in better health.

The continuing multidisciplinary National Child Development Study recruited 17,414 children born in Britain during one week in March 1958 (Centre for Longitudinal Studies http://www.cls.ioe.ac.uk/). When Hitchfield (1978) studied a sample of the brightest children, selected by multiple criteria, she found that in spite of the whole population sampling those measured as gifted were largely drawn from the middle-class. They were also "more stable and less unsettled and maladjusted than the birthweek children as a whole" (p.24) – though their parents worried more about them. One of the later studies, using male data only (a common practice of the time) looked at how the boys' intelligence, measured at the age of 11, was related to their lives at the age of 42 (Nettle, 2003). In Britain, which was becoming much more socially mobile in the late sixties, a boy's high intelligence was found to provide the means to a social status higher than his father's; which would not have been true a generation before.

In Warsaw in 1974, a population cohort of 13,000 11 year-olds were tested for intelligence and school achievement (Firkowska-Mankiewicz, 2002). The subject's achievement levels were about as equally related to their IQs as to their parents' educations, indeed a similar result to these relationships found in "more traditional industrial societies". In this case, though, the research was carried out during a time of "egalitarian social policy", that is before the fall of Communism. Of the high-IQ group, 90% had received secondary education and by their 30s many were in the professions. But of the low-IQ group, only two youngsters had managed education beyond primary level (not quite the proportion in "traditional societies"), both being the children of professionals; and one of them obtained a PhD!

Extracognitive influences

Yet intelligence, however defined and measured, is only part of the complex dynamics of exceptionally high-level performance, which must include extracognitve dynamics such as self-esteem, support and motivation – as well as opportunity (Dweck, 1999; Shavinina & Ferrari, 2004; Freeman, 2004). Barab & Plucker (2002), picking up Vigotsky's (1978) (unreferenced) ideas of the social context of learning, take it further by arguing that perception and cognition are not properties of the individual, but of an environmental transaction, such that talent is an opportunity available to all via "smart contexts" - although it may be actualized more frequently by some. Biometric studies, involving families, twins and adoptees, provide reliable evidence of the environmental and genetic origins of developmental differences, both general and specific (Plomin, DeFries, McClearn & McGuff, 2001). Measurable hormonal differences for the gifted have also been claimed (Ostatníková, 2004).

In the Scottish study (above), childhood intelligence was not always related to how people perceived their success in life (Deary, Whiteman, Starr, Whalley & Fox, 2004). The most reliable predictor in their early years was found to be positive self-esteem, and the most useful tools for actually climbing the career ladder were optimism and pugnacity, similar to what Moon (2002) calls Personal Talent which she describes as teachable. Indeed, Trost (2000), investigating prediction of giftedness in adult life calculated that less than half of "what makes excellence" can be accounted for by measurements and observations in childhood: for intelligence not more than 30%. The key to success, he wrote, lies in the individual's dedication. Others have suggested optimism as the key (Seligman, 1991; Peterson, 2000; Ryan & Deci, 2000).

Work for more than ten years at the John Hopkins University Centre for Talented Youth (CTY) has found that even by the age of 12 the students, who were all volunteers, were significantly different from the general population on the personality test, the Myers Briggs Indicator (Mills, 1993). The most consistent finding was that the majority of the gifted scored highly on intuition, as indeed is claimed for Nobel Prize winners (Shavinina & Ferrari 2004). Mills interpreted this as a preference for abstract and theoretical thinking, whereas most non-gifted students prefer to be factual and pragmatic.

Some follow-up studies are very small. In Australia, for 20 years Gross (2004) has followed up 10 boys and five girls originally aged 11 to 13, chosen because their Stanford-Binet IQs were more than 160. In general the youngsters were found to have low self-esteem, "moderate to severe levels of depression", not to mention "loneliness, social isolation and bitter unhappiness" (p.199) which Gross lays at the door of inappropriate school education. As there were no controlled comparisons with other children it is difficult to tell whether the subjects were representative of other Australian high-IQ children. Of the six American boy "prodigies" followed-up for 10-years, none continued their advantage into adult achievement, a feature of hot-housed children (Feldman with Goldsmith, 1986). Child case-studies provide richness but can miss the wider environmental influences, whether of society or within the family the different interactions of parents with siblings. Each member of a family reacts personally to expectations and encouragement: the outcome being influenced by their genetic, developmental and social perspectives (Freeman, 2000a; Ronald, Spinath & Plomin, 2002; Rutter, 2005).

A 15-year Chinese study of 115 extremely high-IQ children showed the strong influence of family provision, both in achievement and emotional development (Zha, 1995). The children were first identified by parents then validated as gifted by a psychologist. Every year the parents were interviewed several times. By the age of three many children could recognise 2000 Chinese characters, and at four many could not only read well, but also wrote compositions and poems. However, these 'hothoused' children were found to lack easy social relationships so the parents were given lessons in how to help their children get on with others.

There are many concerns about the emotional effects and expectations of the label 'gifted'. When emotional disturbance is associated in a stereotyped way with gifts and talents, and so anticipated by parents, it is more frequently found. Culturally, whereas some children are permitted to be recognised as gifted and talented, that is those who fit the current description, others (e.g. minorities, the disabled and the socially awkward) may not be (Freeman, 2003; Freeman, in press). Parents who use the term gifted have been found to be more achievement-oriented and diminish their children's emotional expression, typically producing less well adjusted children than the parents who did not use this term (Cornell & Grossberg, 1989; Freeman, 2001).

The Freeman Follow-up Study

In 1974, a controlled comparison study began of recognised gifted, unrecognised gifted and random ability children in Britain. The initial concern was to find out why some children were labelled as gifted while others – of identical measured ability and achievement – were not so described. The investigation has used psychological testing and in-depth interviews with the young people, their parents and their teachers in their school and home environments.

The Target group was 70 children aged between five and 14, described as gifted by their parents, almost entirely without testing, all of whom had joined the National Association for Gifted Children (UK). Each Target child was matched with two Control children in the same school class. The First Control was measured as identically gifted though not labelled as such and the Second Control was taken at random (N=210). Each of these triads had also been matched for age and gender. They not only shared their educational experience but were also drawn from the same social population. As there was no discernable difference in the achievements or measured abilities between the Target and First Control children, the essential difference between them was whether or not they had been labelled as gifted.

The whole sample was tested for IQ (Stanford-Binet and Raven's Matrices), personality (Cattell's tests), school behaviour (Stott Behaviour Adjustment Guides) and music (specially constructed tests). Class teachers reported their school achievements and the head teachers of the 63 schools described the school ethos. The audiotaped transcriptions were rated, and together with other data produced 229 variables, which were statistically analysed with orthogonal comparisons and non-parametric analyses. The interview transcriptions were also carefully scrutinised for further information which may not have appeared in the original ratings.

The uniqueness of this investigation was in the methodology of using carefully matched controls, in addition to the long deep interviews over more than three decades. Of the whole sample, 170 children were at the 99th percentile of the Raven's Matrices. Stanford-Binet IQs ranged from the 46 children with less than IQ120 to 18 children with above IQ160; 13 reached the Stanford-Binet test ceiling of 170 IQ. Calculations to increase this quotient do not appear to be either reliable or meaningful. Family finances ranged from very poor to very rich.

Unexpectedly, the audio recordings demonstrated the unreliability of memory, such as when the same incident was described by children and parents separately, even shortly afterwards, or when as adults the subjects remembered their youth. For example, one woman grade-skipped by three years had spent a great deal of her time at Oxford University young, lonely and in tears, but 20 years later remembered that time as blissful. The researcher did not disillusion her. The police are familiar with memory distortion, but researchers and biographical writers seem strangely unaware of it.

There has been attrition over the years so that the 2005 sample, (which is still under search), will probably not be more than 100 subjects. Fortunately, the original groupings are emerging in the same proportions so that outcomes are systematic and recognisable though not yet analysed statistically.

Some findings from the Freeman Follow-up Study

The label

As children, the labelled gifted were usually treated differently by their parents and teachers, whether positively or negatively, and naturally they were aware of adult expectations. Parents might tell me, for example, that their child was too clever to play with others of the same age – in front of the child – and the child indeed found it difficult to have friends. The forces spurring the gifted on to greater advancement could be strong; several subjects rose to the challenge, obtaining doctorates in their early twenties.

Others, though, felt they could never live up to the expectations of giftedness and became big fishes in small ponds, as Zeidner & Scheyler (1999) have described. Typically this would be a student of an extremely high IQ choosing a small college where their cultivated gifted image could shine unchallenged. Some largely ignored their gifts, following their parents into low-level work. Others never managed to fit comfortably into the cut and thrust of work, eventually settling for modest but secure jobs supervised by others. As so many others have found, precocity, extremely high IQ scores and school grades, as well as grade-skipping have not been a route to grown-up high achievements for this sample - except perhaps for those who continued in a similar path, becoming teachers and academics.

As children, the labelled gifted had a far higher incidence of emotional problems ($p<1\%$) when compared with the unlabelled equally gifted. Although the labelled and unlabelled were in the same school class and so experienced identical teaching, parents of the labelled children made significantly ($p<1\%$) more complaints about school provision. The long parental interviews in their own homes disclosed that the labelled gifted children with emotional difficulties had significantly ($p<1\%$) more problematic domestic circumstances, such as parental divorce or experiences which would disturb most children. Their parents would often describe how their child fitted the stereotype of the gifted child, notably the poor peer relationships, and expressed fear for their futures.

Influences on success in life

The most successful adults had been more robust and sociable as children, especially when they had had an external support system of responsive schools, sometimes sincerely felt religion and well above-average intelligence. In terms of conventional success in life such as high examination marks, rising up the corporate ladder or making money, the primary building blocks were always keenness and hard work, allied with sufficient ability, educational opportunity and an emotionally supportive home. High level creativity as seen in adult careers has demanded a particular type of personality which is relatively independent of others' opinions. The successful gifted architect who was a regular school truant, for example, did not do well in his exams and did not show his talents until long after he left university.

Whether conventional and rule-abiding or constantly straining at the leash, the children have usually carried their personal style through to adulthood. Maybe there were no tortured geniuses in this sample, because poor home circumstances, such as a constant change of "uncles" did nothing but harm to the possibility of adult success. In general, it was true that poverty disables while wealth enables. The very rich highly gifted girl, for example, took her first degree at Harvard University as her rightful and natural progression before entering Cambridge University for her higher degree. She now has a prestigious position at the Foreign Office while being the main financial support for her two children and her husband, a physical training instructor. Yet too many of the sample had accepted their parents' views that some of the good things in life such as a professional career were not for them, even though they had the ability to do almost anything they could imagine, and more besides. They opted for modestly-paid work and called it coming to terms with reality.

A clear warning against too much academic pressure emerged from the research. Some seemed to subdue their personalities while striving for high grades, so that their healthy emotional development, including the freedom to play and be creative, was severely curtailed. Such pressure could have the opposite effect from what was intended. The worst affected were the accelerated boys specialising in science who missed out on social relationships. Most of them now say they regret the loss of their childhoods. Of the 17 subjects who had been grade-skipped, only one has fared better in the long-term and several fared less well than had been expected.

There are, of course, many non-scholastic routes to satisfaction in achievement, such as the woman of IQ 170 who was described in the research as empathetically gifted as a child. Throughout school she was effectively the class counsellor, the one to whom the others brought their troubles. She gained a psychology degree and further qualifications and in 2005 cares with love and deep satisfaction for the down-and-outs of her city, being neither well paid nor recognisably a high-flyer.

Most subjects with an exceptionally high IQ, whether labelled or unlabelled as gifted, did much better in life then those with an average score, whatever their original socio-economic level. The most successful had found personal ways of organising their powerful mental abilities: they were more aware and made more efficient use of their personal learning styles. This not only helped them in examinations, but they could elaborate on their learning and take it into adult life. Most high achievers in adulthood have enjoyed a mutually rewarding situation both at home and school, a feeling of comfort with their desire to learn, based on their parents' early pride and support as individuals. The less successful, even with high IQs had remained with less mature and less efficient, shorter-term techniques, like rote-memorising their lesson-notes.

Sometimes far too much of the young people's energy had gone into fighting their school regimes and their teachers, supposedly there to help them. Too many had dissipated their time and energies into wrong channels because of poor educational guidance. At times, the youngsters told me that they knew exactly what they wanted to do, but were thwarted by reasons of school time-table or teacher opinion, and went into areas for which they were less well suited. One girl at a high-powered school, for example, was told that biology was not for her. Defying the school's advice, she secretly entered a competition with her own biological research and won. Only then did the school recognise her potential and permit her to study in the subject area of her choice. She is now a research pharmacist.

The social pressures which can diminish a growing child's feelings of worth were not helped much by the schools and universities they attended. For example, there was neither adequate preparation from her school, nor support from Oxford University for the gentle sensitive girl of IQ 170 from a financially poor family. The social hurdles were too much for her and she soon left to take a modest but emotionally secure job. Of course, no institution should have the power to direct the lives of its students, but without some help, especially for those whose home cannot provide it, the final link in a delicate situation can be lost.

Emotionally, because of the precise ability comparisons, it was possible to see that no matter how 'profound' the gifts and how inadequate the education the gifted had no more emotional problems than others in the same circumstances. Family dynamics made a big difference to the emotional stability of all the children.

Over the years, their happiness and achievements were seen to emerge from each one's personal interaction with social influences.

To support the development of gifted potential most effectively throughout life, it is important to follow indicators such as personal interests (Hany, 1996). Using children's precocity as the prime identifying feature of gifts and talents, with the expectation that precocity would last, could be responsible for their later apparent loss, often called 'burn out'. This may be due to age-peers catching up or the gifted losing interest in the area of their exceptionality. Drawing on my own and others' work, I propose the following system of keeping the door to opportunity open for all giving giftedness a chance to develop in its own manner and time.

Freeman's Sports Approach

Freeman's Sports Approach works on identification by provision, by providing a "smart context for learning". It advocates that given the opportunity and with some guidance the highly able and motivated (features recognised as essential for building excellence) should be able to select themselves to work at any subject at a more advanced and broader level (Freeman, 2000b). In the same way as those who are talented and motivated in sports can select themselves for extra tuition and practice, they could opt for less popular areas such as extra foreign languages or physics. This would mean, of course, that such facilities must be available to all – as sport is – rather than only to those pre-selected by tests, experts or money. It is an inclusive formula. This is neither an expensive route, nor does it risk emotional distress to the children by removing them from the company of their friends. It makes use of research-based understanding of high ability, notably the benefit of focusing on a defined area of the pupil's interest as well as providing each one with what they need to learn with.

To practice the Sports Approach teachers need training in differentiated teaching methods, in addition to a variety of specific techniques for bringing out high-level potential, such as helping pupils to collect information for a portfolio. Most importantly, education authorities should coordinate and share the approach and facilities. Recognition of gifts and talent in this way would also include recognition of the provision to which the students had access. This could be done by a rating scale so that children who were excelling within their context would be seen to be doing so and not penalised because they had poorer provision than others to teaching and material to learn with. An overview of the Approach is presented below.

Freeman's Sports Approach

- Identification should be process-based and continuous
- Identification should be by multiple criteria, including provision for learning and outcome
- Indicators should be validated for each course of action and provision
- The pupil's abilities should be presented as a profile rather than a single figure
- Increasingly sharper criteria should be employed at subsequent learning stages
- Recognition should be given to attitudes possibly affected by outside influences such as culture and gender
- The pupils must be involved in educational decision making, notably in areas of their own interest

After innumerable hours of interaction and investigation with the individuals in this sample as they grew to adulthood, I had to conclude that many influences on happiness and success in life are like love – it is possible to say how it feels and what happens because of it, but there is no sure recipe. For the rest we do have very clear information about what the gifted and talented need by way of support towards adult excellence and self-fulfilment– an education to suit their potential, opportunities to flourish and people who believe in them.

References

Arnold, K. D. & Subotnik, R. F. (1994), 'Lessons from contemporary longitudinal studies', in R. F. Subotnik, & K.D. Arnold, (Eds.) *Beyond*

Terman: contemporary longitudinal studies of giftedness and talent 437-451). New Jersey: Ablex Publishing.

Arnold, K. D. (1995), Lives of Promise: What Becomes of High School Valedictorians? San Francisco: Jossey-Bass.

Barab, S. A. & Plucker, J. A. (2002), 'Smart people or smart contexts? Cognition, ability, talent development in an age of situated approaches to knowing and learning', *Educational Psychologist, 37,* 165-182.

Comford Boyes, L., Reid, I, Brain, K. & Wilson, J. (March 2004) *Accelerated Learning: a Literature Survey.* Unpublished report; Department for Education and Skills, UK.

Cornell, D. G. & Grossberg, I. N. (1989), 'Parent use of the term 'gifted': Correlates with family environment and child adjustment', *Journal for the Education of the Gifted, 123,* 218-230.

Deary, I. J., Whiteman, M. C., Starr, J. M., Whalley, L. J. & Fox, H. C. (2004).The impact of childhood intelligence on later life. *Journal of Personality and Social Psychology, 86,* 130-147.

Dweck, C. S. (1999). Self Theories, their Role in Motivation, Personality and Development. Philadelphia: Psychology Press.

Feldman, D. H. with Goldsmith L. T. (1986) Nature's *Gambit: Child Prodigies and the Development of Human Potential.* New York: Basic Books.

Firkowska-Mankiewicz, A. (2002). *Intelligence and Success in Life.* Warsaw, Poland: IFiS Publishers.

Freeman, J. (1998) *Educating the Very Able: Current International Research.* London: The Stationery Office. (www.JoanFreeman.com)

Freeman, J. (2000a) 'Families, the essential context for gifts and talents', in K. A. Heller, F. J. Monks, R. Sternberg & R. Subotnik, *International Handbook of Research and Development of Giftedness and Talent.* Oxford: Pergamon Press. (669-683) (www.JoanFreeman.com)

Freeman, J. (2000b) 'Teaching for talent: lessons from the research', in Lieshout, C.F.M. & Heymans, P.G. (Eds.) Developing *Talent Across the Lifespan.* (pp. 231-248) London: Psychology Press.

Freeman, J. (2001) *Gifted Children Grown Up.* London: David Fulton Publishers.

Freeman, J. (2002) International *Out-of-school Education for the Gifted and Talented.* Report for the Department of Education and Skills, London. (www.JoanFreeman.com)

Freeman, J. (2003), 'Gender differences in gifted achievement in Britain and the USA', *Gifted Child Quarterly, 47,* 202-211.

Freeman, J. (2004), 'Cultural influences on gifted gender achievement', *High Ability Studies, 15*, 7-24.

Freeman, J. (2005) 'Permission to be gifted: how conceptions of giftedness can change lives', in R. Sternberg and J. Davidson, *Conceptions of Giftedness*, Cambridge: Cambridge University Press.

Gottfried, A. W., Gottfried, A. E. & Guerin, D. W. (in press) The Fullerton Longitudinal Study: A Long-Term Investigation of Intellectual and Motivational Giftedness. *Journal for the Education of the Gifted.*

Gross, M. (2004). *Exceptionally Gifted Children.* London: RoutledgeFalmer.

Hany, E.A. (1996). 'How leisure activities correspond to the development of creative achievement: insights from a study of highly intelligent individuals', *High Ability Studies, 7,* 65-82.

Hitchfield, E. (1978). In Search of Promise: A Long Term Natural Study of Able Children and their Families, Atlantic Highlands, New Jersey: Humanities Press.

Holahan, C. K. & Sears, R. R. (1995). *The Gifted Group in Later Maturity.* Stanford, CA: Stanford University Press.

Mills, C. J. (1993) 'Personality, learning style and cognitive style profiles of mathematically talented students', *European Journal for High Ability, 4,* 70-85.

Moon, S. M. (2002), 'Developing personal talent', Paper presented at the 8th Annual Conference of the European Council for High Ability (ECHA), Greece.

Neisser, U., Boodoo, G., Bouchard, T. J., Boykin, A. W., Brody, N., Ceci, S. J. (1996). 'Intelligence: Knowns and unknowns'. *American Psychologist, 51*, 77-101.

Nettle, D. (2003). 'Intelligence and class mobility in the British population'. *British Journal of Psychology, 94*, 551–561.

Ostatníková, D. (2004), 'Internal environment and the role of hormones in the development of natural abilities', *High Ability Studies, 15, 163-164.*

Perleth, C. & Heller, K. A. (1994). 'The Munich longitudinal study of giftedness', in R.F. Subotnik, & K.D. Arnold, (Eds.) *Beyond Terman: contemporary longitudinal studies of giftedness and talent.* New Jersey: Ablex Publishing.

Peterson, C. (2000). The future of optimism. *American Psychologist, 55*(1), 44-55.

Plomin, R., DeFries, J. C., McClearn, G. E. & McGuff, N. F. (2001). *Behavioral Genetics* (4th edn). New York: W H. Freeman.

Ronald, A., Spinath, F. M., & Plomin, R. (2002) 'The aetiology of high cognitive ability in early childhood'. *High Ability Studies, 13*, 103-114.

Rutter, M. (2005), 'How the environment affects mental health', *British Journal of Psychiatry', 186,* 4-6.

Ryan, R. M., & Deci, E. L. (2000). Intrinsic and extrinsic motivations: Classic definitions and new directions. *Contemporary Educational Psychology, 25*, 54-67.

Schaie, K. W. (2005). *Developmental influences on adult intelligence: The Seattle Longitudinal Study.* New York: Oxford University Press.

Seligman, M. E. P. (1991). *Learned optimism.* New York: Alfred A. Knopf.

Shavinina, L. V. & Ferrari, M. (Eds.) (2004). *Beyond Knowledge. Extracognitive Aspects of Developing High Ability.* Mahwah, New Jersey: Erlbaum.

Subotnik, R., Kassan, L., Summers, E. & Wasser, A. (1993). *Genius Revisited: High IQ Children Grow Up.* New Jersey: Ablex.

Terman, L. M. (1925-1929). *Genetic Studies of Genius* Vols I-V, Stanford: Stanford University Press.

Trost, G. (2000), 'Prediction of excellence in school, university and work', in K. A. Heller, F. J. Mönks, R. Sternberg & R. Subotnik, *International Handbook of Research and Development of Giftedness and Talent.* Oxford: Pergamon Press.

Vigotsky, L. S. (1978). *Mind in Society. The Development of Higher Psychological Processes.* Cambridge, Mass: MIT Press.

Whalley, L. J. & Deary, I. J. (2001). Longitudinal cohort study of childhood IQ and survival up to age 76. *British Medical Journal, 322*, 1-5.

White, K. R. (1992), 'The relation between socio-economic status and academic achievement'. *Psychological Bulletin, 91,* 461-481.

Zeidner, M. & Scheyler, E. (1999), 'The big-fish-little-pond effect for academic self-concept, test anxiety and school grades in gifted children.' *Contemporary Educational Psychology, 24,* 305-329.

Zha, Z. (1995), 'The influence of family education on gifted children'. Paper presented at World Conference on Gifted and Talented Children, Hong Kong.

Spearman's *g*: Past, Present, and Future[1]

Douglas K. Detterman

Spearman's *g*: Past, Present and Future

In this paper, I will review the past, present and future of Spearman's *g*. I will begin with a brief history of the development of the concept of *g*, or general intelligence. I will then discuss what I think is presently well known about general intelligence and will support these conclusions with specific research studies. Although these findings are well known within the community of those who study general intelligence, they are not as widely known outside of that community. These facts should be more widely known because they have very important implications for society. When I consider the future of *g*, I will consider two things. First, I will discuss the important questions that still need to be answered. And, second, I will consider how what we already know could lead to important social issues that will need to be resolved. Throughout this paper, the opinions expressed and the speculations are mine and mine alone. However, I believe it is probable that a large portion of those who study intelligence would endorse what I have said here.

Past

The first question is why there is an interest in intelligence at all. An examination of history shows very little interest in intelligence before 1800. Most of the few mentions have to do with determining mental competency or differentiating handicapped

---

1 Author Note: Parts of this work were supported by grant No. HD07176 from the National Institute of Child Health and Human Development, Office of Mental Retardation. Requests for reprints may be sent to D. K. Detterman, Department of Psychology, Case Western Reserve University, Cleveland, OH 44106. E-mail: Detterman@case.edu

persons (e.g., discriminating mentally ill from mentally retarded) for legal purposes. Not surprisingly, the increased interest in intelligence is most closely related to the increase in literacy and the formal education necessary to acquire literacy.

From the Reformation onward, there was an increased emphasis on literacy. One of the chief arguments of reformers was that religious thought should be directly available to everyone, not just the church elite. This led to a movement to teach everyone to read religious texts, initially in Latin and eventually in their own language. Reading and writing were not always taught together. For example, for nearly 200 years in Sweden most people were able to read but not write.

The need to learn to read and write led to the development of formal schooling and it was in these situations where differences in cognitive ability were apparent. One of the first scientists to take an active interest in the differences among people was Francis Galton. One of his first efforts was 'Hereditary Genius' first published in 1869. In this book, Galton traced the lineage of English aristocracy across generations. He compared the success of families to each other and also to sons adopted by the Popes through history, as was their custom. He concluded that some English families showed consistent proof of high accomplishment while others were less accomplished. The adopted sons of Popes generally did not achieve what would be expected of someone raised with all of the advantages of the aristocracy. Since the Popes' sons served as a rough control for environment, Galton concluded that the observed differences among the English aristocratic families must be due to hereditary differences.

It is surprising that Galton was able to make this observation. It must be remembered that the mechanism for genetic transmission of characteristics was unknown at the time. However, the practical skill of improving farm animals through selective breeding was widely practiced and it was well known that some traits could be passed on. There were even popular books around this time that told people how to select the appropriate mate supposedly based on the principles of selective breeding. Most of these were pure nonsense but they do show that people of that era realized that traits could be genetically transmitted.

It is in 'Hereditary Genius' that Galton gives what I believe is one of the first and perhaps the best descriptions of general intelligence:

> People lay too much stress on apparent specialties, thinking over rashly that because a man is devoted to some pursuit, he could not possibly have succeeded in anything else. They might just as well say that because a youth had fallen desperately in love with a brunette, he could not possibly have fallen in love with a blonde. He may or may not have more natural liking for the former type of beauty than the latter, but it is as probable as not that the affair was mainly or wholly due to a general amourousness of disposition. It is just the same with special pursuits. A gifted man is often capricious and fickle before he selects his occupation, but when it has been chosen he devotes himself to it with a truly passionate ardour. After a man of genius has selected his hobby, and so adapted himself to it as to seem unfitted for any other occupation in life and to be possessed of but one special aptitude, I often notice, with admiration, how well he bears himself when circumstances suddenly thrust him into a strange position. He will display an insight into new conditions, and a power of dealing with them, with which even his most intimate friends were unprepared to accredit him. Many a presumptuous fool has mistaken indifference and neglect for incapacity; and in trying to throw a man of genius on ground where he was unprepared for attack, has himself received a most severe and unexpected fall. I am sure that no one who has had the privilege of mixing in the society of the abler men of any great capital, or who is acquainted with the biographies of the heroes of history, can doubt the existence of grand human animals, of natures pre-eminently noble, of individuals born to be kings of men. (Galton, 1869, pp. 24-25)

It is hard to imagine a clearer literary statement of what *g* is. It is a general ability, like a general amorousness, that can be applied to anything. It is also surprising how much a person high in general intelligence knows about even in areas outside those to which they have devoted themselves.

'Hereditary Genius' made an impression on no less a figure than Charles Darwin. In a letter on December 23, 1869 Darwin wrote the following to Galton: "You have made a convert of an opponent in one sense, for I have always maintained that, excepting fools, men did not differ much in intellect, only in zeal and hard work; and I still think this is an eminently important difference (Darwin Correspondence Project, #7032)." What is most interesting about this letter is that Darwin, who had already made his reputation on his astute observations and interpretations of physical variations in animals, had to be convinced by Galton that cognitive differences exist among people.

I do not find it surprising that Darwin had to be convinced. I have heard present day psychologists express the same sentiment. I am frequently amazed at how impervious people are to noticing individual differences in cognitive ability, even people whose job it is to study human behavior. There may be two reasons for this. First, it may be that we live in a *g*-segregated world. Second, it is much more satisfying to attribute what we have accomplished to our own hard work rather than the luck of the genetic draw.

Galton was never able to mathematically formalize his observations on *g* because he had not yet invented the statistics necessary for the job. That task fell to a second year English graduate student (already in his middle years after leaving a military career) working in the Leipzig laboratory of Wundt. Thirty-five years after Galton's observations, Spearman (1904) published what is arguably the most important paper that has yet been published in psychology. In this paper, he presented a formalized notion of general intelligence and also a mathematical test of the position. The mathematics ultimately led to the development of factor analysis, a technique essential for understanding *g*.

While a full development of *g* is beyond the scope of this paper, the basic concept that Spearman presented is easy to grasp. He noted that all mental tests were correlated with each other (positive manifold). He suggested that all of the tests were positively correlated because they all measured general intelligence or *g*. Tests measured different amounts of *g*. For example, if one test was composed .8 of *g* and another was composed .4 of *g* then the two tests would correlate .32 (.8 X .4 = .32), the product of what is now known as their *g* loading. Of course, in the real world no one knows what the g-loading is of a test. Working backwards from the correlations between tests it is possible to determine the *g*-loading of each test and the amount of g for each subject using factor scores.

Over the last hundred years, the story has become more elaborated with much more sophisticated statistical methods, with the addition of other secondary factors less prominent than g, and with much better tests. However, the essential point that Spearman made has not changed and, in fact, has only been reinforced by a hundred years of research findings. Today *g* is, without question, the most powerful psychological construct as judged by predictive validity, research support, or nearly any other criterion.

People are sometimes confused about what *g* is. *g* is what, in psychology, is called an unobserved variable or a latent trait. There are many latent variables in psychology. They include anxiety, hunger, and depression to name just a few. Because we can identify a latent variable doesn't mean that we know what it is. For example, though we have a test to measure depression, it doesn't mean that we know what depression is or, even, that depression is a thing that actually exists as a single entity.

Latent variables are not peculiar to psychology. Gravity, the g from physics, is a latent variable. Though there is no doubt that there is gravity and that it can be measured, no one knows exactly what it is. Obviously, we would like to know exactly what gravity is just as we would like to know what general cognitive ability is.

This thumbnail description of the history of *g* has only presented the major developments. It is, however, amazingly complete in that it presents the essential details. What has been left out is primarily the supporting research that makes *g* such an important construct and will ultimately help us understand what *g* is.

Present: What Do We Know?

In this section, I will present what is currently known about g. This is research that I believe is widely supported within the field to the extent that there is no longer major debate on these issues. For some of the issues to be discussed, the support is overwhelming while for others it is just strong. There is not space here to present an exhaustive review of the support for these conclusions so I will present an example of recent research that makes the case. For each of the points discussed, it should be remembered that the research is much more extensive than what is cited here.

g is amazingly uniform across test batteries.

If you take a battery of, say, 8 tests and use that battery to identify *g* for each person who takes the battery and then give the same group of people another battery of 10 tests and again identify *g* for each person in the group, what will the correlation be between *g* from the first battery and *g* from the second battery? It is hard to believe that two completely different batteries of tests would result in identical measures of the same construct. But this is what we must expect of any valid scientific construct. If each scale used gave a different and unpredictable measure of weight, we could not use weight to develop a scientific construct.

Over the years, this issue has been resolved for general cognitive ability. There have been sufficient demonstrations that the *g*'s obtained from different batteries are equivalent. This is sometimes referred to as indifference of the indicator, meaning that it doesn't matter what the indicator of *g* is. The limitation of these studies has generally been how much testing people could tolerate. Fortunately, there is a study that has justification for testing people to the point of exhaustion. The Minnesota Study of Twins Reared Apart (MISTRA) consists of twins who were raised apart and their relations. This is such a rare and important sample that it justified extensive cognitive, personality, and medical testing.

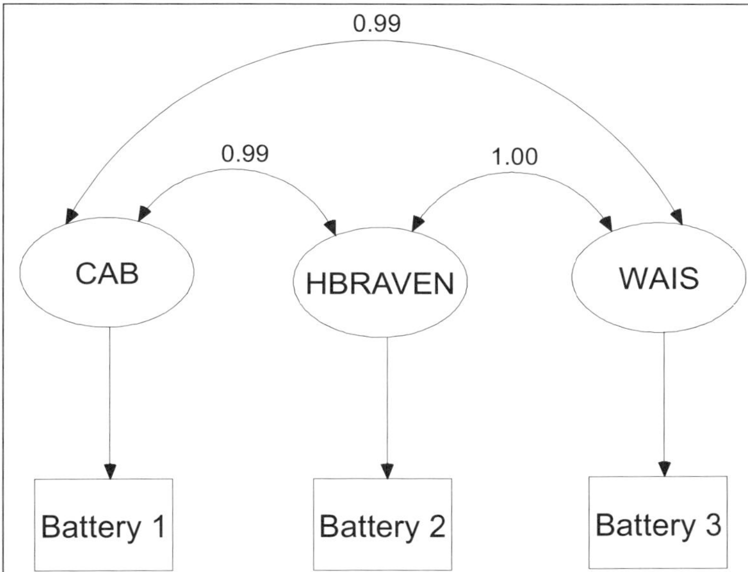

Figure 1. A diagrammatic representation of the results of Johnson, Bouchard, Krueger, McGue, and Gottesman (2004) showing almost perfect agreement between the *g*'s extracted from three different batteries of cognitive tests.

Johnson, Bouchard, Krueger, McGue, and Gottesman (2004) recently analyzed the three batteries that were given to 436 participants in MISTRA to determine the extent to which the *g*'s extracted from each battery would agree. The batteries contained a wide range of tests and all the tests were different. Each battery also had a slightly different emphasis on what mental abilities were being tested. Appropriate subfactors and a general factor were extracted from each battery. The correlations between each of the *g*s extracted were then computed. Figure 1 shows a diagrammatic representation of the results. As can be seen from the diagram, two of the three possible correlations were perfect (1.00) while the third was very close to perfect (0.99).

The conclusion, then, is that *g* obtained from any representative battery of tests will be the same as *g* obtained from any other representative battery of tests. It should also be noted that when batteries yield a single IQ score, the IQ score is generally so highly correlated with *g* as to be indistinguishable from it. Even when investigators don't go to the trouble of extracting a *g* score for each subject, conclusions regarding the global IQ score from the battery are usually the same as they would be had the analysis been based on *g*. In the remainder of this paper the terms *g*, general cognitive ability and intelligence will be considered to be interchangeable.

It should also be pointed out that *g* is perhaps the one area in psychology where multiple batteries using different tests would yield results as consistent as this. It seems highly unlikely, for example, that multiple tests of any personality trait or disorder would yield results as consistent as those that are obtained for *g*. It is doubtful that there are any other latent traits in the social sciences that even approach to the kind of invariance across methods of measurement that *g* shows.

g is stable over age within subjects.

Another characteristic of general cognitive ability is its stability over time for an individual. From the very beginning it was well known that IQ tests were generally highly reliable. The best individually administered tests have reliabilities at or above .90. There have also been a number of demonstrations of reliability across fairly long periods of time.

Perhaps the ultimate study of stability (reliability) of IQ was recently conducted by Deary, Whalley, Lemmon, Crawford, and Starr (2000). It so happens that all Scottish children born in 1921 were tested on the Moray House Test in 1932 at age 11. Deary happened to wander on to these data late in the 20th century. He and his coinvestigators located 101 subjects from the original sample in 1998, making the participants 77 years of age at the time they were relocated. Deary et al readministered the Moray House Test to the group under the same conditions. (The Moray House Tests are actually a set of tests used for educational selection purposes heavily weighted with verbal reasoning measures.)

The obvious question is how similarly those tested on the same test they had taken 66 years earlier would perform. The astounding result is that the correlation, when corrected for attenuation in range, was .73. This is a reliability that is higher than the immediate test-retest reliability of many psychological and medical tests. I cannot think of any human characteristic that would evidence this kind of stability from age 11 to age 77. It is doubtful that even physical characteristics like height, weight, blood pressure, pulse rate, or anything else would show this kind of reliability over this period of time.

Not only is general cognitive ability stable but changes in ability appear to be genetically programmed. The classic work of Wilson showed that identical twins frequently showed the same variations in mental development.

General intelligence is heritable (h^2 = .40-.80)

There has been no more contentious issue than if general intelligence is a heritable trait. That question has been settled by the mass of evidence in support the heritability of general intelligence. Over the years there have been several hundred studies on over forty-thousand persons that have reported the correlation between participants with various degrees of relationship. The more closely related people are, the more highly similar they should be in general cognitive ability and so the higher the correlations should be among relatives that are more closely related.

Bouchard and McGue (1981) thoroughly summarized the studies that have been carried out. As Figure 2 shows quite clearly, the more closely genetically related people are, the more similar they are in general cognitive ability. The figure shows that as kinship is reduced so is the correlation between kin. Monozygotic twins are most similar in general cognitive ability and the correlation is nearly as high for monozygotic twins raised apart as it is if they are raised together. Of course, monozygotic twins are genetically identical and if general cognitive ability was perfectly heritable, the correlation between them would be perfect. While it is not perfect, it is above .70. Some regard monozygotic twins raised apart as the best single estimate of heritability.

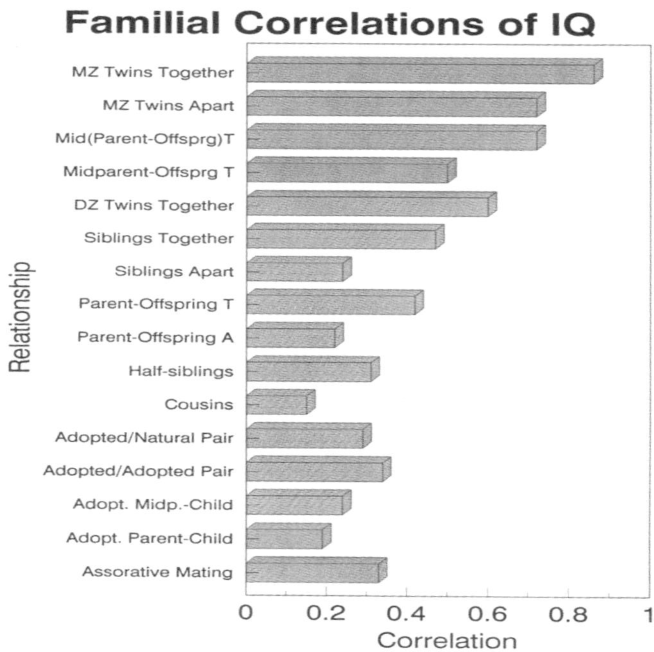

Figure 2. The correlation of general cognitive ability among persons of different degrees of relationship. Based on data from Bouchard and McGue (1981).

Other entries in Figure 2 show the magnitude of correlation that would be expected if intelligence was heritable. While it is possible to estimate heritability in several ways, the most thorough estimates are obtained if all the data are used simultaneously. This has been done by a number of investigators including Chipauer, Rovine, and Plomin (1989) using modelling procedures that allow the use of all data in estimating heritability and environmentality. The results from this and other studies are summarized in Figure 3. Heritability accounts for 51% of total variance. Common environment, those environmental influences that individuals share, accounts for 22% of total variance. Specific environment, those environmental influences unique to a particular persons, account for 27% of common environment.

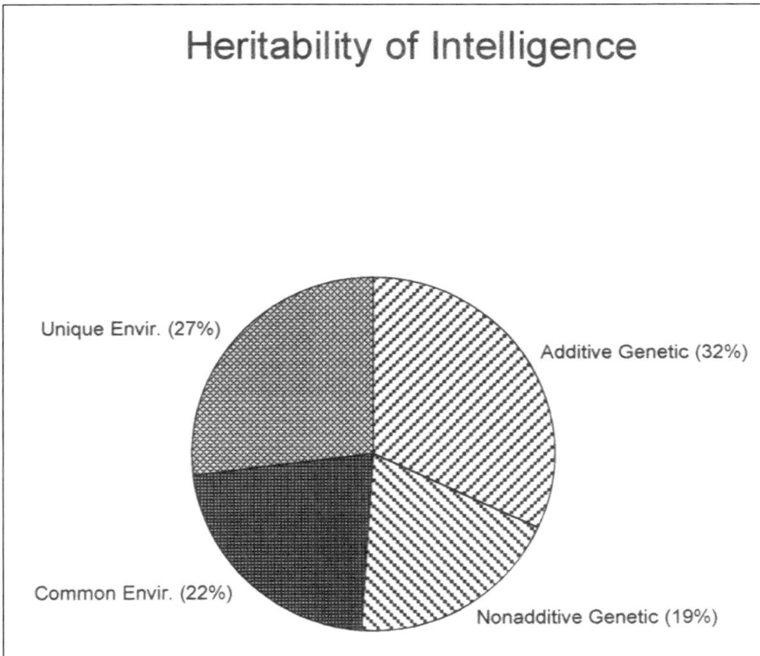

Figure 3. Best estimates of hereditary and environmental contributions to intelligence.

In summary, based on these data it would appear that heredity and environment each account for about 50% of general intellectual ability. The heritability of general intellectual ability is so well established that some well known behavior geneticists have suggested that no more behavioral genetic studies are necessary to establish the heritability of general intellectual ability.

The heritability of g changes with age.

Studies of heritability where all the data available are entered into a single model provide an excellent summary of heritability and environmentality for all of the studies that have been conducted.

However, they could be averaging studies over significant variables that could make a difference in the results.

It must be remembered that heritability is not a fixed, immutable constant. As conditions change, heritability also changes. For example, if it were possible to optimize environments so that environment was perfect for each person, environment would no longer be important in producing differences among people. Since environment would no longer be important in accounting for differences, all differences among people would he due to heredity and heritability would be 1.0. Genetics would account for all of the differences among people.

One variable that has been investigated for its affects on heritability is age. Many studies using behavioral genetic designs have been done with children because they are often easier to identify and test. There is good reason to believe that age could have an impact on heritability. It is known that general intelligence correlates most highly with years of school completed later in life and that, when a person has two careers, the later one is usually more representative of their general cognitive ability.

When studies are sorted by age, the proportion of variance in general cognitive ability due to common environment, specific environment and heredity are seen to change with age. The most important general trend is that the influence of common environment seems to begin fading in adolescence until it fades to nearly zero by early adulthood. This is, perhaps, not surprising since children are heavily influenced by peers throughout adolescence and generally leave their home in later adolescence. What is surprising is that rearing environment has so little lasting effect on cognitive ability, although this should have been predictable from studies of intervention (see below).

The effect of specific environment also declines slightly with age. It is difficult to tell if this decline is actual or an artifact. The problem is that the portion of total variance due to test unreliability is included in specific environment and tests become more reliable as children grow older. Therefore, it is difficult to know if specific environment effects are actually changing or tests are becoming more reliable. To separate these two portions of variance would require correction for reliability of tests over age.

Since both common and specific environment decline as children get older, heritability must increase. The proportion of variance accounted for by genetic factors for children less than 6 years old may be below .40. However, by early adulthood heritability increases to between .55 and .85.

***g* is difficult to change through active intervention.**

Another well established fact that comes as a surprise to those not familiar with research on general cognitive ability is that general intelligence is very difficult to change if it can be changed at all. There have been a large number of intervention studies, some of which have been extremely ambitious. Spitz (1986) reviewed all of these and concluded that there was little or no evidence that any of them had a positive impact on IQ.

Perhaps one of the best done and most extensive interventions was carried out in the Abecedarian Project. Subjects were randomly assigned to groups and the 27 subjects in the control group received 40 hours a week of intervention from shortly after birth until they began school. Both the experimental group and the 28 subjects in the control group received nutritional, medical, and dental interventions. While the intervention was being administered, it appeared that it was working quite well. Unfortunately, after a few years in elementary school, the effects of intervention appeared to largely "fade out". Later follow-ups suggested about a four point superiority in IQ of the experimental group over the control group. There is some suggestion that randomization may not have worked and that the experimental group was at least this different from the control group from the beginning. Nevertheless, of the many intervention studies that have been conducted this is the best and most extensive. If the gain of 4 points is not an artefact, it is a very small gain for the effort.

One problem with intervention studies is just that, they are interventions not permanent changes in an environment. It seems logical that if someone from a bad environment is put into a good environment and then put back in a bad environment, their IQ would initially increase but that this increase would "fade out" when they were returned to the bad environment. Producing permanent changes in general cognitive ability may require that there be permanent changes in environment.

There are few, if any, identified environmental variables with substantial influences on g.

Many investigators have claimed to identify environmental variables that affect intelligence. These include variables like parenting style, books in the home, vocabulary parents' use, and socioeconomic status. Most of these claims are based on a simple correlation of some variable that is 'obviously' an environmental variable with *g*. The problem is that no variable is obviously an environmental variable. If there are a lot of books in the home, it could be that the parents have high IQs and high incomes, like to read, and buy a lot of books. Socioeconomic status has also frequently been identified as an index of environment. However, most indexes of socioeconomic status include variables like parental occupation, income and others that are either directly or indirectly related to the heritable portion of *g*, In short, identification of 'obvious' environmental variables almost always confounds genetic and environmental sources of variation.

The only certain way to identify an environmental variable is to use a behavior genetic design where variance can be divided into genetic and environmental components. In the cases where this has been done, the results have been disappointing. Usually few variables can be identified and when they are they only account for a small portion of the variance.

Why is it so difficult to find environmental variables that affect general cognitive ability? There is no doubt that environment affects general intelligence. From the previous discussion of IQ, we know that environmental variance accounts for somewhere between 20% and 50% of the variance in *g*. Flynn (1987, 1998) has pointed out a substantial gain in IQ that has taken place over time now called the Flynn effect. The cause for this gain is not known.

One reason that environmental effects may be so difficulty to detect is that they may be either small and idiosyncratic or so pervasive that they affect every one about equally. By adulthood, it appears that most of the effects due to common environmental variance have been lost and all that remains is the unique environmental variance. Given the varied paths that individuals take through life, it may be very difficult to identify exactly what the unique influences were that directed their lives. In fact, it may not be a single event that channels a person's life but a confluence of small events.

On the other hand, something like the Flynn effect may be caused by a very broad environmental change that affects everyone in the population. For example, there has been a general improvement in nutrition over the last hundred years for a variety of reasons. Such a change is apparent in the increase in height that has occurred but would be much more difficult to detect in general cognitive ability because of measurement characteristics.

***g* differs across intelligence level**.

Detterman and Daniel (1989) divided the IQ distribution into five equal parts and then examined the correlation within each segment of the IQ distribution. They found that those with the lowest IQ had the highest average correlations among mental tests. The average correlation among tests decreased as subjects' IQ increased.

This is a very odd finding because it indicates that the effect of *g* is highest in persons with low IQs. Those with the highest IQ will have the lowest correlation among mental tests suggesting that intelligence may be less of a 'general disposition' in this group. I think this finding is most interesting, though, because it sets a restriction on theories of *g*. Any successful theory will not only have to explain how *g* arises but show how different levels of correlation occur across IQ levels.

Summary: What we know.

The evidence is overwhelming that *g* is heritable, stable, and difficult to change. No one has been able to identify environmental variables that affect *g* or to change *g* through systematic intervention. This is a much different picture of general intelligence than most people have, particularly if they depend on the popular press for their information about g. It may also be a different picture than many people want to have. For better or worse, that is the way the truth is revealing itself.

Future: Critical Questions

Now that I have presented my view of the current status of what is known about *g*, two things may be apparent. First, there are a number of very important unanswered questions and, second, what we already know about *g* has implications for a number of important social issues. I begin by listing the most important unanswered questions about *g*.

What is *g*?

One question that has been skirted in this discussion is exactly what g is and the answer is that nobody knows for sure though there has been a great deal of speculation. *g* is well identified by statistical operations and there is no doubt about its scientific validity. However, there is not a well articulated, rational theory about how *g* arises.

To explain *g*, we must explain why all mental test results are correlated with each other, as Spearman observed over 100 years ago. Why, if someone has a large vocabulary, will they also be good at solving spatial reasoning problems. (You may substitute any two mental tests for vocabulary and spatial reasoning in the previous sentence.) To paraphrase Galton, how do we explain this general disposition to be smart.

One of the first thoughts that come to most people is that *g* is probably explained by some single property of the brain that unifies mental operations. For example, neural conduction speed may differ to produce differences in *g*. A single property would certainly be the most parsimonious explanation of *g*. Such an explanation would require that there be a very high correlation between a single variable like conduction speed and *g*. Unfortunately, no variable has been found that is consistently highly correlated with *g*. Therefore, it seems unlikely that *g* can be related to a single thing though it is still a possibility if someone can find something that is consistently highly correlated with *g*. To be convincing, the correlations will have to be nearly as high as the reliability of intelligence tests, over .90. Another problem is that it is hard to understand how a single variable would produce higher correlations at lower intelligence levels.

Another way of explaining *g* is by invoking multiple processes like processing speed, working memory, and perceptual identification. The multiple processes, whatever they may be, must, in combination, predict *g*. Detterman, Mayer, Caruso, Legree, Conners, and Taylor (1992) were able to show that a set of basic cognitive variables was able to predict general intelligence as well as intelligence tests predict each other. The one difficulty with this approach is that it cannot predict different levels of correlation among tests at different IQ levels as was found by Detterman and Daniel (1989).

Detterman (1987, 1994) suggested a system approach that can account for both differences in levels of performance as well as differences in correlations across IQ levels. The basic concept is that *g* is composed of a number of independent cognitive processes. Mental activities generally require that several of these processes be used. The processes are arranged in an operational system and some of the processes are central to the operation of the system. A central process is, as the name implies, one that is crucial to the operation of the entire system and will be used for all mental activities.

Figure 4 shows a model of a very basic system. Inputs occur at A, B, and C and outputs at G, H, and I. This system has 9 potential processes that can be identified by tracing a path from A, B, or C to G, H, or I. ADEFG and BDEFI are two of the nine processes in this simple system. As can be seen, D, E, and F are included in every chain of operations and are central processes.

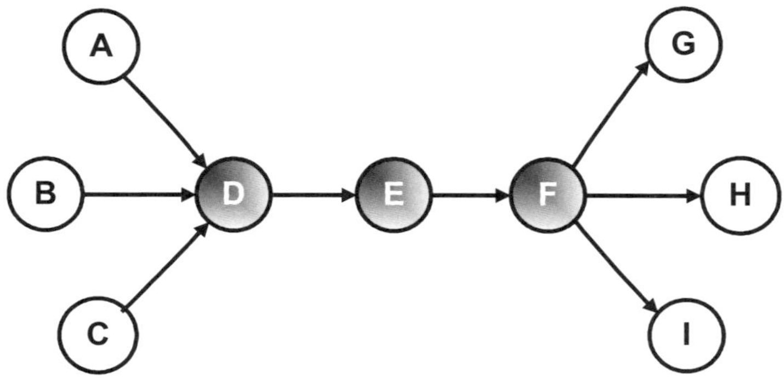

Figure 4. A model explaining general intelligence based on deficits in central processing.

Since all processes are independent, deficits can occur in any process. The level of system operation for any single mental activity will be set by the weakest process in the chain. Deficits that occur in central processes are particularly bad because they affect all operations that the system performs and cause a general decrement in the system efficiency. Persons with a deficit in a central process will perform more poorly on all tasks they are asked to perform. A deficit in a central process acts as the 'weakest link' or 'road block' in the system and sets the level of performance for the whole system.

For those persons who have adequately functioning central processes, the efficiency of the system will be determined by the weakest peripheral process. How efficiently each of the mental processes or pathways operate will be more variable than if there had been a large deficit in a central process. What this means is that for persons with larger central process deficits, all processes will function more poorly and, so, be more highly correlated with each other. On the other hand, people with very efficient central processes will do better overall but the level of performance will be set by the weakest peripheral process. This means that their performance across mental processes will be more variable and, so, will have lower correlations among these processes.

Although I quite obviously prefer a system explanation, it is not yet clear how *g* will eventually be explained. There are certainly other theoretical possibilities that have not been considered here. Whatever the final explanation is, I consider understanding the nature of *g* the top research priority. Until *g* is understood, it will be difficult to know how to move forward on other related questions.

What is the biological basis of *g*?

There has been substantial speculation on the biological basis of intelligence. Biological hypotheses about *g* are bound to increase with the large number of new techniques that have become available including Positron Emission Tomography (PET), Magnetic Resonance Imagery (MRI), and Averaged Evoked Potentials (AEP), to name a few. Most of these studies attempt to find those areas of the brain that are most important to *g*.

A study illustrative of this trend is a PET study by Duncan et al. (2000). They asked participants to solve items that had previously been shown, in a psychometric study, to be either higher or lower in *g* loadings. PET imaging was then used to record brain activity. By subtracting the activation that occurred during low *g* loaded stimuli from those that occurred during high, the researchers were left with an image of those brain areas most active during the solution of the high *g* items. The results suggested frontal areas of the brain were most closely related to *g* and the investigators concluded that they had found the "neural basis for *g*". Other studies have supported this conclusion, some using MRI methods.

I find at least two problems with these conclusions and fear that they may be premature. First, it is not clear to me exactly what imaging techniques are showing. The fundamental control would be to compare a brain thinking to a brain not thinking. I have a great deal of difficulty imagining what a brain not thinking would be. It very well could be that the most important processes involved in general intelligence are those that are always maximally engaged in the awake brain. Keeping the brain turned on would seem to me to be an optimally adaptive strategy. If the techniques being used are missing the processes that occur at a constant level, then frontal lobe activity might be necessary but not sufficient to explain *g*.

A second problem is that other sources of information would suggest that the cortex including the frontal lobes are not that important in accounting for general intelligence. For example, studies of soldiers who received bullet wounds through the cortex show little effect on intelligence with some studies showing no effects. This is hard to reconcile with the frontal cortex being the neural basis of general intelligence.

What is the genetic basis of *g*?

Understanding the genetic basis of *g* is as critical as understanding its biological or behavioral roots. The ideal explanation of *g* would trace it from genes to brain to behavior. Such an understanding would also require a full description of how environment affects each level. It should be understood that what is being discussed here is the genetic mechanisms that affect *g*: how individual genes affect the brain and its efficiency. From the previous discussion it should be clear that there is no doubt that there are genetic influences on *g*. But we only know this at the behavior genetic level based on correlations among relatives. That is different from knowing explicitly what genes affect intelligence.

Plomin and Spinath (2004) have provided a good summary of the quest for understanding the genetics of intelligence. Several investigators have been attempting to identify genes associated with intelligence. None of these efforts has been as successful as might have been predicted, though a number of genes have been identified that are related to educational disabilities. Even though the techniques for searching for genes have become quite sophisticated at maximizing power it is possible that they may not be powerful enough. If, for example, there were 300 genes associated with *g* and each accounted for only a third of one percent of the variance, it will take extremely powerful techniques to identify them and methods presently being used would not be sufficient.

Another possibility it that genes may not have simple effects on IQ. Some have suggested that mutation load may be a possible explanation for *g*.

What implications does g have for education?

What we know about *g* is on a collision course with what we are learning about education. This is happening at every level of education from preschool to university. I will cover each of these in turn.

In the US, Headstart began in 1965 as an early intervention program designed to increase economically deprived children's intelligence and readiness for school. Initially it was a summer program that cost about $96 million and served 561,000 children. It has expanded into a full year program that now serves about 900,000 children and costs over $6 billion dollars per year. As a basis for comparison, in 2004 the US spent about $95 billion on all medical research. In 40 years, the expenditure on this program has grown more than 625 fold while enrolment has not even doubled.

Perhaps the worst part is that the program does not meet the original goals set for it. It increases neither intelligence nor academic achievement (Caruso, Taylor, & Detterman, 1982). According to its supporters, the program produces many secondary gains, including making medical and dental care available, socializing for school, etc. However, eventually someone will ask if this program is really cost effective. I cannot help thinking that this money could better be spent on a much more research oriented program that would seek to understand early intellectual development.

The US has recently launched a new initiative called "No Child Left Behind" (NCLB). One aspect of the program is regular, systematic achievement testing of, initially, primary school children and now extending into secondary school. The goal of testing is to provide benchmarks for schools so that they can improve performance, when necessary. Hardly anyone could object to that. Ultimately, the hope is that this program will increase high school graduation rates, equalize achievement across socioeconomic groups, and increase achievement at every level.

While these are all admirable goals, the problem is that we do not understand how to accomplish any of them. Variables typically discussed like amount spent per pupil, class size, teacher effectiveness, and other non-student variables each account for pathetically small amounts of the total variance. What does account for the largest portion of variance in student outcome? Student incoming ability, that is, the general cognitive ability of the incoming student accounts most heavily for outcome. Until we understand general cognitive ability we have little chance of developing optimum educational opportunities.

This is supported by a study by Thompson, Detterman, and Plomin (1991) in which they showed that when achievement was not consistent with general intellectual ability, environment was correlated with the discrepancy. But in cases where achievement and general intellectual ability were the consistent, both were largely controlled by common genetic influences.

Frustration is already beginning to emerge in the NCLB program. Some schools are appealing decisions regarding their tests and what standards they should have to meet. As time goes on, I would expect that anxiety will increase. Administrators of inner city schools are placed in a very difficult position. They are required to improve the performance of their students but no one is telling them how to do it.

One might think that the problem would end at the college level since continuation of an education beyond secondary school is optional. Not so in the US where high school graduates compete to gain admission to the 'best' universities. Admission is heavily based on a standardized test, often the SAT, which, itself, is an IQ test (Frey and Detterman, 2004). Anghoff and Johnson (1988) studied to what extent attending a selective school affected subsequent performance on the Graduate Record Exam (GRE), a test used to select students for graduate school. They located the records of 22,293 students who had taken the SAT and then 3.5 for 4.5 years later, took the GRE. They studied quantitative scores alone because those seemed most affected by their variables of interest. They entered SAT, major, gender, and the interaction of these variables into a regression equation to predict GRE. The total variance accounted for ranged from 93.2% to 99.9% depending on the exact details of how the regression was conducted. This means that, at most, differences in colleges could account for 7% in the total variance in outcome. In Anghoff and Johnson's own words:

> The interpretation of this result is that, for a given level of initial ability, the variation in GRE-quantitative scores between students within an institution swamps the variation in scores between institutions. This might be taken to indicate that, given initial ability, individual (i.e., within-school) characteristics are much more important in determining the final GRE score than are institutional level characteristics. (p. 45)

What they are saying is that there is essentially no difference in what college a person attends. From this work, it is clear that the most important variable in college outcome is the person's general cognitive ability at the time they are admitted. When a few other variables like gender and major are taken into account, there is not much left to account for. For college teachers it is a little discouraging to think that there is only a maximum of 7% of the variance available to work with and that 7% has to be shared with test error variance.

As a general conclusion, I believe it is safe to say that *g* predicts between 60% and 80% of educational outcomes at every level of education and this is probably a low estimate. If this is true, it is obvious that there will not be effective educational intervention programs until we understand *g*. Even if we do have a full understanding of *g*, it is possible that we will not be able to control all of the variance that contributes to *g*. This means that there may never be a day when everyone will achieve to the extent many would desire. In that case, we will have to develop ways of humanely dealing with the variability that does exist.

What implications does *g* have for social policy?

There are a number of other social issues that are very likely to receive substantial public attention. Many of these will be related to the growing underclass. It is almost inevitable that the underclass will increase as society becomes more complex and educational attainment becomes increasingly important for employment. It is important to understand that the increase in the underclass has little to do with race. It surprises most people to find that the average difference between siblings is about 11 IQ points. This can be determined empirically by comparing the absolute IQ difference between siblings. For the Western Reserve Twin study, the mean absolute difference between same-sex dizygotic twins is 10.2 points. The differences can also be computed theoretically as the variance of IQ multiplied by the correlation of sibling IQs. The square root of this quantity will give you the standard deviation of the distribution of the absolute difference on IQ scores for sibilings. The mean absolute value of the difference between all pairs in a normal distribution can be obtained by multiplying the standard deviation of the distribution by 1.1. If we assume that the standard deviation of IQ is 15 and the correlation of IQ between siblings is .45, the result for the theoretical calculation is 11.06. What this means is that there is almost as much variation between siblings as there is between races. Even if there were no race differences, there is enough within family variation to assure a large underclass.

So what issues will be important concerning intelligence and social policy? Gottfredson (2004) has already documented the case for *g* as the cause of social class inequities in health care. Persons with lower IQs are less able to follow directions, less likely to seek medical help, less able to describe symptoms, and have many other barriers to optimum medical care.

Although the case has not been as well made as for health care it might be expected that future research will show that persons with a low IQ are less capable of proper birth control planning, seeking legal assistance, doing proper financial planning, dealing with government agencies to get help, finding appropriate job training opportunities to get a better job, and on and on. Over the next few years I expect to see an explosion, led by Gottfredson's work, of research documenting the difficulties the cognitive underclass has.

Once these difficulties are documented, the next question is what social policies can and should be developed. While most people who are scientists actively avoid speculating on social policy issues, speculation may not only be called for, it may be unavoidable. It is to be hoped that those who do enter this arena do not do it on the basis of idle speculation but can justify what they suggest with hard facts and models that indicate probable outcomes. So far, I have seen no suggestions about social policy that meet the criteria of being based on hard facts and having predictable outcomes.

Conclusions

This tour de force has attempted to provide an overview of the history, present knowledge and hoped for future directions for research and social policy concerning g. Much of what I have presented is either my own opinion or my speculation. My guess is that many in the field would share my views and there are many who would consider them understated. Unfortunately, I do not think that either the public or the popular press or even our colleagues who do not study intelligence are fully aware of much of what I have said or its implications. There seems to be substantial resistance to coming to a clear understanding about *g* and its implications. If you don't believe me, read almost any introductory psychology text chapter on intelligence. These chapters avoid mentioning much and in some cases even deny much of what I claim is well known and agreed upon.

While science must always be willing to entertain every reasonable hypothesis, those who propose hypotheses that are in opposition to accumulated data like those presented here must present data that support their position. If they do not do that, they cannot be regarded as credible. Perhaps we in the field will have to tell people when the king is naked rather than remaining politely silent.

References:

Anghoff, W. H. and Johnson, E. G. (1988). *A study of the differential impact of curriculum on aptitude test scores.* Research Report pp. 88-46. Princeton, NJ: Educational Testing Service.

Bouchard, T. J., Jr., and McGue, M. (1981). Familial studies of intelligence: A review. *Science*, **212**, pp. 1055-1058.

Caruso, D. R., Taylor, J., and Detterman, D. K. (1982). Intelligence research and intelligent policy. In D. K. Detterman and R. J. Sternberg (Eds.), *How and how much can intelligence be increased* pp. 45-65, Norwood, NJ: Ablex Publishing Corporation.

Chipauer, H. M., Rovine, M. J., and Plomin, R. (1989). LISREL modelling: Genetic and environmental influences on IQ revisited. *Intelligence*, **14**, pp.11-29.

Darwin Correspondence Project, Cambridge University Press, www.lib.cam.ac.uk/Departments/Darwin/.

Deary, I. J., Whalley, L. J., Lemmon, H., Crawford, J. R., and Starr, J. M. 2000). The stability of individual differences in mental ability from childhood to old age: Follow-up of the 1932 Scottish Mental Survey. *Intelligence*, **28**, pp.49-55.

Detterman, D. K. (1987). Theoretical notions of intelligence and mental retardation. *American Journal of Mental Deficiency*, **92**, pp.2-11.

Detterman, D. K. (1994). A system theory of intelligence. In D. K. Detterman (Ed.). *Current topics in human intelligence: Vol. 4. Theories of Intelligence*, pp. 85-115. Norwood, NJ: Ablex Publishing Corporation.

Detterman, D. K. (1994). Intelligence and the brain. In P. A. Vernon (Ed.). *Handbook of the neuropsychology of individual differences,* pp 35-57. San Diego, CA: Academic Press.

Detterman, D. K., Mayer, J. D., Caruso, D. R., Legree, P. J., Conners, F., and Taylor, R. (1992). Assessment of basic cognitive abilities in relation to cognitive deficits. *American Journal of Mental Retardation*, **97**, pp.251-286.

Detterman, D. K. & Daniel, M. H. (1989). Correlations of mental tests with each other and with cognitive variables are highest for low IQ groups, *Intelligence*, **13**, pp. 349-359.

Duncan, J., Seitz, R. J., Kolodny, J., Bor, D., Herzog, H., Ahmed, A., Newell, F. N. and Emsle, H. (2000). A neural basis for general intelligence. *Science*, **289**, pp. 457-460.

Flynn, J. R. (1987). Massive gains in 14 nations: What IQ tests really measure. *Psychological Bulletin,* **101**, pp. 171-191.

Flynn, J. R. (1998). IQ gains over time: Toward finding the causes. In U. Neisser (Ed.), *The rising curve: Long-term gains in IQ and related measures,* pp. 25-66. Washington, DC: American Psychological Association.

Frank, S. A. and Slatkin, M. (1992). Fisher's fundamental theorem of natural selection. *Trends in Ecology and Evolution*, **7,** pp. 92-95.

Frey, M. C. & Detterman, D. K. (2004). Scholastic assessment or g? The relationship between the SAT and general cognitive ability. *Psychological Science,* **15**, pp. 373-378.

Galton, F. (1870). *Hereditary genius: an inquiry into its laws and consequences.* New York: D. Appleton & Co.

Gottfredson, L. S. (2004). Intelligence: is it the pidemiologists' elusive "fundamental cause" of social class inequalities in health? *Journal of Personality and Social Psychology*, **86**, pp. 174-199.

Johnson, W., Bouchard, T. J., Jr., Krueger, R. F., McGue, M., Gottesman, I. I. (2004). Just one *g*: consistent results from three test batteries. *Intelligence*, **32**, pp. 95-107.

Plomin, R., & Spinath, F. M. (2004). Intelligence: Genetics, genes, and genomics. *Journal of Personality and Social Psychology*, **86**, pp. 112-129.

Spearman, C. 1904. "General Intelligence" objectively deteremined and measured. *American Journal of Psychology*, **15**, pp. 201-293.

Spitz, H. (1986). *The raising of intelligence: A selected history of attempts to raise retarded intelligence.* Hillsdale, NJ: Lawrence Erlbaum Associates.

Thompson, L. A., Detterman, D. K. and Plomin, R. (1991). Associations between cognitive abilities and scholastic achievement: Genetic overlap but environmental differences. *Psychological Science*, **2**, pp.158-165.

Wilson, R. S. (1972). Twins: Early mental development. *Science*, **175**, pp. 914-917.

Taking Social Intelligence Seriously[1]

R. I. M. Dunbar

Introduction

For the past 150 years, the emphasis in the study of individual differences has focussed on what we might refer to as instrumental intelligence – the ability to solve memory or reasoning puzzles that relate principally to the everyday physical world as we experience it. This has, of course, been coloured to varying degrees by the fragmentation of IQ into its sub-components (verbal reasoning, arithmetical ability, etc), and taken to more abstract depths in the work on general intelligence (*g*). I shall argue here that much of this work has overlooked an important component of intelligence, namely the suggestion that the human brain (or mind) has been adapted principally to solve problems associated with the social world rather than the physical world. This being so, we may need to ask what its implications for intelligence testing might be.

First, however, there is an important issue as to whether or not social intelligence is a specialised module (in the sense sometimes advocated by some – but not all! – evolutionary psychologists). I shall argue here that social intelligence is not a specialised module as such, but rather an emergent property of more conventional cognitive processes, and hence that aspects of intelligence like "*g*" (the so-called "general intelligence" factor) are in fact central to social intelligence. This has important implications for how we understand the nature of intelligence in general, and *g* in particular.

---

[1] British Academy Centenary Research Project, School of Biological Sciences, University of Liverpool, Liverpool L69 7ZB

What is Social Intelligence?

Humans, like most mammals, are intensely social. In many ways, primates' success from an evolutionary perspective is a direct consequence of that sociality. Primate societies are implicit social contracts which allow some of the problems of survival and reproduction to be solved cooperatively. Social contracts of this kind work because they allow relevant problems to be solved more efficiently. However, social contracts require individuals to be willing to forego some of their more immediate personal interests in order to benefit from greater returns later through group-level cooperation. If too many individuals act in their own selfish interests, the cohesion of the group will be threatened simply because too many others will end up paying the costs of sociality. Group stability is quickly threatened, leading to the rapid collapse of the contract.

The real issue here seems to be the cognitive demands of maintaining the stability of relationships through time. That process of negotiation is far from simple. At its most basic, it requires co-ordination and compromise, and often subtle adjustments to ensure that group members do not drift apart during foraging. But it also requires the ability to manage conflict in such a way as to reduce its natural tendency to cause groups to dissipate, and to cope effectively with the ever-present threat generated by rivals. In the limit, and certainly in the human case, it requires individuals to be able to understand another's perspective sufficiently well to appreciate what kinds of adjustments are necessary to create the levels of "bondingness" required to keep a group together. One element of that is knowing when to trust another individual.

However, the very fact that these processes depend on relatively sophisticated cognitive skills leaves them open to abuse. Freeriders, who take the benefits of sociality without paying all the costs, are an inevitable by-product of a biological system that is built on cognitive flexibility rather than genetic hardwiring or endocrinological determinism (as is the case with many social insects, for example). Freeriders destabilise the social contract because they oblige others to pick up the bill. If that bill becomes too heavy, it will wipe out the benefits that these individuals expect to accrue from sociality and their willingness to cooperate in the contract will soon evaporate.

Although the case for social cognition is well established in principle, we do not have a very clear idea about what is involved. We know much about only one aspect of social cognition, and this is the phenomenon known as "theory of mind" or ToM. And even then, our knowledge is limited mainly to its natural history and we know almost nothing about the phenomenon itself as a cognitive process (Roth & Lesley 1998). Theory of mind is the ability to understand another's mental state, to understand that another individual believes something to be the case. It forms a component of what philosophers refer to as "intentionality", the state of mind associated by terms like believing, intending, supposing, desiring and so on. In this context, ToM is identified with second order intentionality ("I believe that you intend [...to do something]"). First order intentionality is equivalent to having knowledge of one's own mind state ("I believe [...that something is the case]"), and is thought to be the condition for most – if not all – higher vertebrates, although it could well be more widespread.

Second order intentionality (theory of mind itself) is a major transition point in child development and occurs at around age 4 years as a very discrete phase change in a child's understanding of the social world (Astington 1993). The transition is important because it allows children to appreciate that another individual may hold a belief about the world which is different to that held by the child (a so-called "false belief"). At this point, it is able to factor another individual's belief state into its calculation of events. This has an important consequence for the child's capacity to interact with the social world in which it is embedded because it facilitates two activities that it could not previously do: it can engage in fictional play (pretending that the doll can drink tea) and it can lie (deliberately make statements about the condition of the world that it knows to be untrue in order to mislead another individual and cause them to behave in a way that will benefit the child).

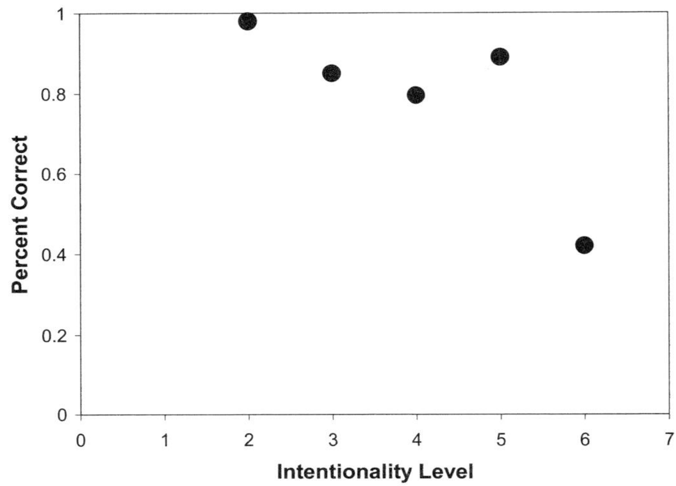

Fig. 1. Normal adults' performance on advanced mindreading tasks, as a function of the number of levels of intentionality involved. Subjects were read short vignettes of a character thinking about future plans that involved the likely mental states of other individuals, and were then asked to respond "true/false" to statements about the mind states of the various characters. The subject's beliefs count as level 1. Subjects' performance at levels 2 to 5 did not differ significantly from 100%, but that at level 6 did. After Barrett et al (2002).

The philosopher's notion of intentionality is implicitly reflexive: in principle, intentionality forms an embedded hierarchy of mind states that has no end. "I intend that you suppose that I want you to understand that I believe that you… " – the sequence is limited only by the human mind's ingenuity in constructing meaningful sentences reflecting the mind states of those involved. However, it should be self-evident that the human mind cannot manage to keep track of an unlimited number of these mind states, and in practice the limit seems to occur at fifth order (Fig. 1) (Kinderman et al 1998, Stiller & Dunbar submitted). Even so, this is quite impressive, given that 4-year-olds can master only two orders and most animals (with the possible exception of great apes) only one.

I will have more to say about the cognitive mechanisms underpinning intentionality later, but it is worth pointing out here that there is some evidence to suggest that this form of social cognition seems to be relatively expensive in wetware terms. With the very limited data we have available, there is some evidence that achievable levels of intentionality across species of primates are correlated with neocortex volume, and specifically frontal lobe volume (Dunbar 2003a). As we shall see, this in itself is interesting, but the main lesson for the moment is that the computational demands of intentionality (and, more generally, social cognition) seem to be very heavy, and this may in turn provide us with a possible candidate for the selection pressures that have driven brain evolution in primates in particular, and perhaps mammals in general. This has given rise to what has become known as the "social brain hypothesis" and it is to a more detailed consideration of this that I now turn.

The Social Brain Hypothesis

For a long time, the reason why primates have so much larger brains for body size than all other vertebrates (Jerison 1977) remained somewhat unclear, although the presumption inevitably was that this must have to do with the kinds of ecological strategies that different species pursued. In due course, a number of alternatives were put on offer (Table 1). These included the need to navigate around the home range (species with larger ranges presumably needed larger brains to handle this), the nature of the foods on which they feed (fruits are usually more patchily dispersed than leaves, so frugivores need larger brains to track the ephemeral food sources on which they depend) and extractive foraging (foods that are embedded in some kind of matrix from which they have to be extracted will be cognitively more demanding).

Table 1. Hypotheses that have been proposed over the years to explain the evolution of brain size in primates.

Hypothesis	Explanatory focus	Core references
Ecology	frugivory (ephemeral food sources)	Clutton-Brock & Harvey (1980)
	range size (mental maps)	Clutton-Brock & Harvey (1980)
	extractive foraging (embedded foods)	Gibson (1990)
Social learning	innovations (social transmission)	Laland & Reader (2002)
Social skills	Machiavellian behaviour	Byrne & Whiten (1988)
	social bonding	Dunbar (1992, 1998), Barton (1996)
Developmental	energetic constraints	Martin (1981)
	structural constraints	Finlay & Darlington (1995)

In addition, a number of non-functional hypotheses suggested that large brains were essentially a by-product of developmental processes. One suggested that larger bodied animals gained savings of scale in terms of energy consumption that could be invested in disproportionately large brains. The other suggested that larger brains were simply consequences of allometric scaling relationships dictating brain growth. While both these hypotheses may well be true, and have their own intrinsic interest, they are not functional explanations. Since brain tissue is amongst the most expensive in the body (Aiello & Wheeler 1995), they do not explain why a species that might not need a large brain should invest so heavily in one. Doing so would be evolutionarily counter-productive and would be heavily selected against. We still need to provide a functional explanation that offers us an explanation in terms of selection pressures.

The problem with all the ecological hypotheses listed above is that, while offering a clear selective advantage to large brains, they fail to explain why species like primates should need brains that are so much larger than those of other mammals, such as ungulates (the deer-antelope group) or even carnivores. However, during the 1980s, Byrne & Whiten (1988) proposed that the answer lay in the unusual complexity and sophistication of the primate social world. They couched this in terms of the use of tactical deception and coalition formation, and referred to their hypothesis as the "Machiavellian Intelligence Hypothesis". The suggestion aroused some considerable interest, but there was relatively little empirical evidence to support the hypothesis, whereas there was at least *prima facie* evidence to support most of the ecological hypotheses.

During the 1990s, it proved possible to adduce considerable empirical support for the Byrne-Whiten proposal. The principal evidence derived from demonstrating that social group size correlated with brain volume (and, in particular, neocortex volume) in primates (Dunbar 1992, 1998). In this context, neocortex volume is of particular interest because it is this component of the brain that has enlarged out of all proportion during the course of primate evolution (Finlay & Darlington 1995). Whereas the neocortex accounts for 10-40% of total brain volume for most mammals, in primates it accounts for 50-80% (the latter being in adult modern humans). Thus, when we ask why primates have large brains, what we really mean is: why do they have large neocortices? The neocortex is interesting because, in very crude terms, it is the seat of what might be considered the "smart" elements of cognition (rational, flexible decision making). This is not, of course, to say that other parts of the subcortical brain are not involved in high level cognition, but it is probably to say that some of the more crucial components of those processes are localised in cortical rather than subcortical structures. The most obvious sense in which this is so is that the cortex, and perhaps especially the frontal lobes, play a crucial role in integrating outputs from subcortical and cortical structures, not simply in the sense that the association cortices link perceptual facts together but by imposing meaning onto facts about the world (Frith 1996).

An alternative version of the social brain hypothesis has been offered in recent years, and this is the suggestion that social learning lies at the root of the computational demands (Reader & Laland 2002). The emphasis here is again more strictly ecological, but, like the Byrne-Whiten version, it emphasises that the ecological problem-solving is done socially rather than individually. This suggestion reminds us that, ultimately, all these explanations have an ecological root. The conventional social brain hypothesis itself assumes that the animals are attempting to solve problems of everyday survival or reproduction, but do so socially (i.e. communally as a group) rather than individually, as assumed by the conventional ecological hypotheses.

The distinction between the social learning and the social cohesion versions lies only in that the former specifies a mechanism of information transmission as the fundamental constraint, whereas the latter makes no particular specification as to what cognitive mechanisms are involved. The social learning hypothesis has tended to focus on foraging problems, and it may be that learning how to solve foraging problems more efficiently is an evolutionarily efficient way to behave. But in principle, there are no particular grounds for distinguishing between social and ecological problems as the target of social learning. However, it is worth noting that the social brain hypothesis in its standard (or Machiavellian) form implies that the issue is less about social transmission of learning *per se* than about managing and manipulating large bodies of information on the relationships between members of a social group.

It is possible, for the sake of argument, to create clear water between these two versions of the social brain hypothesis: a distinction can be drawn between whether the information constraint lies in social transmission of behavioural styles (be these social or ecological in focus) or in managing and exploiting information about the behaviour of others so as to ensure the group's cohesion. In one respect, this distinction alludes to the longstanding debate among primatologists as to whether primate sociality owes its origins to the need to maximise foraging return or the need to minimise predation risk (see Dunbar 1988). Since this debate has largely been resolved in favour of predation risk as the driving force for sociality (Dunbar 1988), this contrast between the two versions probably fails at the first hurdle in favour of the social-brain-as-bonding hypothesis. Given this, the more useful way of viewing the Laland-Reader version is either in terms of its emphasis on particular cognitive mechanisms (namely, social transmission of learning) or as a by-product of social cognitive skills evolved to facilitate social bonding.

The principal evidence adduced in support of the social brain hypothesis is that social group size correlates with relative neocortex volume in primates, whereas a selection of indices of ecological constraints do not correlate well with neocortex volume (Dunbar 1992). The finding that social group size correlates with relative neocortex size in primates, and that humans seem to fit into the same quantitative pattern (Dunbar 1993) has since been generalised in two different directions. One has been to show that a whole suite of behavioural (as opposed to strictly demographic) indices also correlate with neocortex size. These have included the size of grooming cliques (essentially, coalitions) (Kudo & Dunbar 2000), male mating strategies (Pawlowski et al 1997), the amount of social (as opposed to object or solitary) play (Lewis 2000), and the amount of tactical deception (Byrne & Corp 2004). This shift of emphasis to actual behavioural strategies was given further support by Joffe's (1997) demonstration that, in primates, neocortex size correlates best with the length of the juvenile period, whereas the brain as a whole correlates best with the period of parental investment (gestation plus lactation). The significance of this is that the juvenile period is the period of socialisation, the period during which the animals' social skills are developed and honed (often, of course, by imitation and social learning). Thus, Joffe's finding offers evidence of an important role for what amounts to software development: the size of the neocortex is important, but part of that importance lies in its capacity to absorb and use information and experience – and that itself is probably a function of the quantity of neural machinery that can be brought to bear on the problem.

The second direction in which these ideas have been developed during the last decade has been to extend the social brain hypothesis to a number of other mammalian orders. There is now robust evidence to suggest that the relationship between brain size and sociality holds in carnivores (Dunbar & Bever 1998, Pérez-Barbería et al submitted) and ungulates (Pérez-Barbería & Gordon 2005, Shultz & Dunbar 2005), with some evidence to suggest that this is probably also true for cetaceans (Morino 1998), bats (Dunbar 1993) and the advanced insectivores (Dunbar & Bever 1998). Although these latter findings point to the generality of the social brain hypothesis, there is important evidence to suggest that the nature of the relationship in primates is subtly different from that which holds more generally across other mammalian taxa. This seems to reflect the fact that the nature of primate sociality is different from that in mammals in general, mainly because it depends on the kinds of intense sociality only found within monogamous pairbonds in other mammals (Shultz & Dunbar, in preparation). In most mammals, this form of bonding probably has a great deal to do with scheduling the activities of individuals so that they coincide sufficiently to maintain group coherence through time and space. But in primates, it seems to extend to the more general problem of integrating the interests of several individuals, who have conflicting interests at stake.

Social Brain and Instrumental Intelligence

The social brain hypothesis emphasises the fact that it is in the social domain that the principal cognitive costs arise, and not in those aspects of everyday life that might be considered the domain of instrumental intelligence. In this, it makes a stark contrast with the assumptions adopted by more conventional approaches to intelligence, which have tended to focus on domains such as verbal fluency, numerosity, inference and to some extent memory that are functionally important in an educational context.

At this point, there would appear to be two ways of resolving this contrast in perspectives. One is to argue a strong claim that conventional intelligence testing is looking at the wrong end of the animal: it is tantamount to focussing on athletic performance as a measure of health. Granted that athletes are healthy people, it does not follow that all healthy people are necessarily athletes. By the same token, focussing on instrumental intelligence may completely fail to address the fact that cognition exists principally to solve complex social problems and has specialisations in this respect. Failing to notice this may mislead us into emphasising some skills at the expense of more important ones, and therefore social cognition should be included as a formal part of the definition of intelligence. The alternative possibility is that social cognition of the kind we have discussed is simply the outward observable consequences of more fundamental, but quite conventional, cognitive processes that are basically just those identified in the intelligence testing literature.

The difference between these two positions lies in the distinction between the modular and generalist approaches to cognition. Some schools of evolutionary psychology might be tempted to argue for the first, and claim that social cognition is a specialised cognitive (or even neuropsychological) module that uses quite different cognitive strategies to those used by instrumental problem-solving. Although these schools have claimed that an evolutionary approach necessarily implies such modularity, in fact this does not necessarily follow. It certainly *may* imply modularity, but evolution is frankly indifferent as to which way the system is organised: whatever way it happens to turn out, that's simply the way evolution has implemented it. Evolution is not *always* efficient in its modes of operation, though it can be. Moreover, we should not assume that modularity is necessarily more efficient than more generalised systems: this is a purely empirical question and it has yet to be resolved.

It is probably important, however, to distinguish between high and low level modularity. This corresponds to the distinction between Fodorian modularity and full-blown modularity of the kind implied by, for example, Pinker (2002). Fodorian modularity corresponds to neurological specialisations that are dedicated to handling specific perceptual cues or similar low-level functions that require automated processing. Modularity at this level may often be an efficient way of organising cognition, but these are less demanding than high level modules of the kind that are sometimes advocated in the evolutionary psychology literature. The important issue for us in this context is that many of the features that constitute the core of conventional intelligence testing look more like Fodorian modules (i.e. basic processing units) and could thus constitute the cognitive underpinnings for a more general yet specialised form of social cognition. It is to this possibility that I now turn.

The Nature of Social Cognition: A Proposal

How, then, are we to articulate the relationship between conventional approaches to intelligence and social cognition? Although we could argue that social cognition represents a specialised neurological module, a more interesting suggestion (interesting in the sense that it offers us more scope for drawing inferences about quantitative differences between species) is the suggestion that social cognition itself is an emergent property of the scale at which more conventional forms of cognition are implemented. This is a strictly computational hypothesis: in other words, it argues that theory of mind and allied forms of social cognition depend in a very explicit way on the scale of core cognitive competences (Barrett et al 2002).

The suggestion that the capacity for theory of mind (in particular) is simply a by-product of underlying cognitive competences has been argued by at least one school of developmental cognitive psychologists (e.g. Mitchell 1997). They have suggested that theory of mind as seen in 5-year-old children is really just a reflection of developing executive function (i.e. those cognitive skills localised in the frontal lobe of the brain – the so-called "executive brain"). It may be no accident, therefore, that children's acquisition of theory of mind seems to correlate with (or, more precisely, follow close on the heels of) the development of many of core cognitive skills (not least, of course, language, but also many of the familiar Piagetian cognitive stages). However, while it may seem obvious that functions like memory are important, the present proposal would identify a specific set of more advanced cognitive skills as being critical: a minimal list would be causal reasoning, analogical reasoning, the capacity to engage in mental rehearsal (planning) and the time frame over which projections can be made (i.e. how far into the future projections about the consequences of actions can be made). The latter, in particular, may well depend on the development of episodic memory (the capacity to place events into a biographical timeframe: Tulving 2002).

Many, or at least most, of these may be widespread cognitive functions across at least the more advanced animal species (i.e. birds and mammals). The issue here is not the functions themselves, but rather the scale at which they can be executed and the extent to which they can be focussed onto a single problem context (e.g. the likely future behaviour of an individual or the choice between two alternative behavioural strategies). More explicitly, we have argued that what we see or experience as theory of mind is simply what happens when these functions can be executed on a sufficiently large scale that it generates the ability to reflect on others' mental states. It is particularly obvious, perhaps, that the ability to understand another mind is in some way dependent on analogical reasoning, since this is precisely how we reason about other's feelings and emotions: we cannot experience other's mental states directly, so we are obliged to do so exclusively by analogy with our own, using external cues like facial expressions and vocalisations to bridge the experiential chasm between our own mind and another's.

If the scale at which these can be executed is a function of the computing power of the brain, and the computing power of the brain is some function of its size (and, hence, neural density and connectedness), then it follows that the scale of social cognition might be a function of brain size. And we may expect this to apply both across species and, within species, over the developmental period of brain growth. In other words, we should not expect to see evidence of theory of mind in children until such time as the brain has developed to sufficient size and organisational stability to cope with the computational demands. It may therefore be no accident that advanced functions like theory of mind do not appear in children until around the point at which brain size is reaching its final adult proportions (i.e. around age 5 years).

It is perhaps worth pointing out in this context that the higher orders of intentionality (3rd to 5th order) that characterise normal human adults develop in stages between 6 and 12 years of age (Barrett et al, submitted). Whether this reflects continuing structural adjustment of the developing brain or whether it reflects socialisation processes that involve learning and experience remains unclear. A case could be made, following Tomasello (2000), that these higher orders appear as a result of an ontogenetic ratchet when children develop in social communities: children who lack social experience and/or the tutoring offered by parents may be unlikely to develop these skills on their own. Indeed, it may be that, in addition to the socialisation ratchet, there is an additional cultural ratchet in which especially insightful individuals are able to train others who might not otherwise have discovered the ability to explore higher orders of intentionality if left to themselves. If so, it remains unclear whether the "natural" (i.e. untutored) limit is at 5th order or at some lower level. Since only a few individuals appear to be able to perform at levels higher than 5th order, the highest realistic limit probably lies at around 7-8th order (Stiller & Dunbar, submitted).

Though we have yet to explore this aspect of social cognition in detail, there is some *prima facie* evidence to suggest that the levels to which individual species can take these skills may be a function of some aspect of brain volume. The most direct evidence for this is the fact that the limiting achievable levels of intentionality for Old World monkeys, apes and adult humans (taken to be 1st, 2nd and 5th order intentionality, respectively) are a linear function of frontal lobe volume (Fig. 2a). Another relevant finding is that the speed with which orang-utans, chimpanzees and human children can solve a standard set of puzzle boxes without opportunities for prior practice or observation is also a linear function of frontal lobe volume (and is not related either to total brain volume or to total neocortex volume) (Fig. 2b). If this relationship holds up, it would offer the first clear evidence of a cognitive constraint imposed by the computational power of the organism's brain. However, more extensive data are needed before we can be prescriptive on this point.

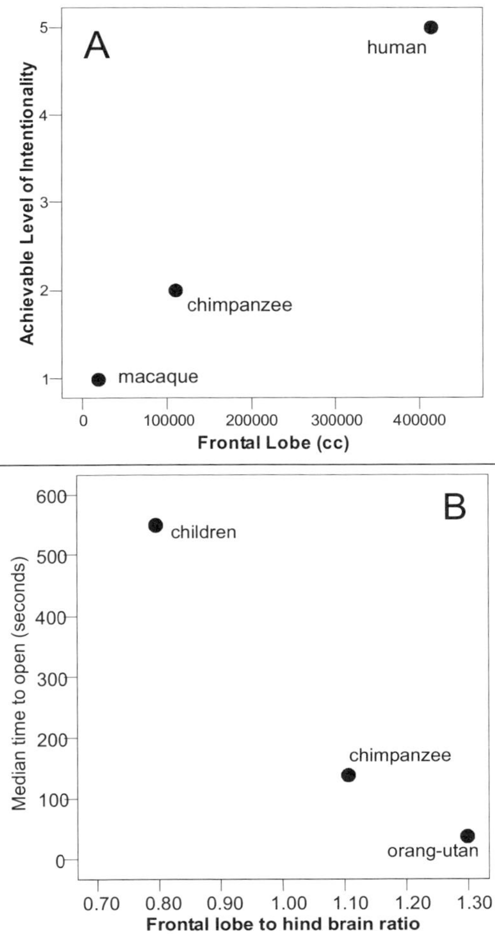

Fig. 2. (a) Upper limits on mindreading abilities of humans, apes and monkeys, based on performance on mindreading tasks. After Dunbar (2003); (b) Median speed with which children and great apes successfully opened puzzle boxes scale linearly with frontal lobe volume. After Dunbar et al (2005).

Nonetheless, if we can take these results at face value, they provide some preliminary evidence to support the suggestion that it is with the frontal lobe (or "executive brain") that these capacities seem to correlate best, thus reinforcing the suggestion that there may be a direct link between social intelligence and more conventional instrumental intelligence (and, more particularly perhaps, *g* itself). That said, there is some obvious clinical evidence to suggest that the correlation might be less than perfect. People with Asperger syndrome, for example, may be of normal (or, occasionally, even high) instrumental intelligence (in terms of performance on conventional intelligence tests) but are socially inadequate. So far as I know, however, there is no evidence to suggest that anyone who scores poorly on conventional intelligence tests might be unusually competent in the social domain. There are, of course, classic syndromes in which individuals are highly sociable yet of limited or low instrumental intelligence (Down syndrome and Williams' syndrome are obvious instances), but the nature of sociality in these cases is in fact always impaired. Such individuals are often overly-trusting in their interactions with others, and often unskilled in the kinds of subtle reading of signals that is necessary for close social engagement.

Specialisations of Social Cognition

I want, in this final section, to suggest that social cognition may have important consequences that do not arise directly out of basic cognitive processes *per se*. These consequences concern two areas of human endeavour and experience that often seem to be overlooked in this context. They are story-telling and religion.

Story-telling and religion have two key features in common: they require us to be able to imagine virtual worlds that have no immediate experiential content and they require us to assume that those whom we ask to share these imaginings with us can genuinely follow us on this journey. I can perhaps show this most easily in the context of drama. Consider Shakespeare sitting down to compose his play *Othello*. His problem is this: to make a play interesting, you need to have at least three characters, and you have to ensure that your audience understands what is going on inside the characters' minds – anything less, and the story becomes a dull narrative. So, the audience must *understand* that Iago *intends* that Othello *believes* that his wife Desdemona *wants* to run off with Cassio (which would probably not be much more than idle fantasy by Desdemona were Iago not able to convince Othello that Cassio himself also *wanted* the same outcome). So stated, it will be obvious that, if they also have to factor Cassio's complicity into the equation to make the deception convincing for Othello, the audience has to be able to work at fifth order intentionality. But to do this, Shakespeare himself must operate at one level higher: he must *intend* that the audience *understands* … etc. Shakespeare was having to work comfortably at sixth order intentionality, and this is now one level beyond the normal limits for most adult humans.

We can look at the play another way that tells much the same story. In practice, of course, the story is played out in a series of scenes in which a variety of characters come and go, and express their feelings and concerns. If we look at how Shakespeare handles the structure of his plays in more detail, then we find that a typical scene involves four speaking characters (Stiller et al 2004). As it happens, this mirrors very closely indeed the limits on natural human conversations (Dunbar et al 1995). So when Shakespeare was constructing the play and putting the story and characters together scene by scene, he was actually working with four characters' mind states, and he was thus expecting his audience to work at fifth order if they were going to be able to follow the convolutions of the plot as it developed. At this point, Shakespeare himself, of course, had to work at sixth order.

One interpretation of this analysis is that the difference between a good story-teller (be they dramatist or novelist) is that they can take their audience to their natural cognitive limits and still make it work. In doing so, of course, it is important that they do not push the audience beyond their natural limit, otherwise the story will become incoherent for them. Indeed, those of Shakespeare's plays that are conventionally regarded as difficult for audiences (*Titus Andronicus* is a notable example) are precisely those in which he tries to include too many characters. So good story-tellers tread a fine line here: they must judge the audience's capacities so that they do not under-challenge them (or the story will seem uninteresting) and at the same time not over-challenge them (else they will lose all but the small number of cognitively very skilled individuals).

What makes story-telling cognitively demanding is that we (audience and composer alike) have to imagine a fictional world. Even though, in a conventional drama, the characters are on stage in front of us, nonetheless we have to make a leap of imagination to suppose that these characters (the actors) are someone other than whom they actually are. To be able to even imagine that possibility, we need second order intentionality (theory of mind). As I have already noted, Barrett et al (2003) have argued that this ability to work in a virtual (or fictional) world is especially demanding cognitively and may explain why great apes (and hence humans) have such large neocortices compared to other primates.

I have suggested elsewhere (Dunbar 2003b, 2005) that the same demands are true of religion. Religion in any meaningful sense requires us to imagine that there is another universe that we cannot directly see and touch: this spirit world exists along side the physical world we experience directly, and has its own reality as well as being able to interact with our world. We can conceive that such a world exists, but to do so must minimally require second order intentionality (theory of mind). I must believe that the world as I experience it is not all there is. Though not necessarily a statement about other minds, we can, I think, argue a case here that believing in two simultaneous but separate worlds requires us to run two belief states in parallel – hence requiring second order intentionality. In effect, I have to believe that something is both true (the world is as I see it) and not true (the world as I see it is not the only possible world) at the same time. So far so good, but to make this interesting as a social phenomenon, I have to believe that you also believe this to be the case. But we still don't have anything approximating religion here: merely a belief in another parallel universe. To make it into something even approximating a real religion, we both have to believe that there are beings (spirits?) in this parallel universe who themselves have intentions [about how we should behave]. At this point, we require fourth order intentionality. Even so, we still do not have religion in the communal sense that seems to me to be fundamental to the nature of religion – i.e. such that it can be used to enforce conformation to the communal will. To do so, we have to add one further layer of intentionality: you and I have to understand that we both believe that these denizens of the spirit world intend that we should behave with appropriate intentions [otherwise they will exact retribution]. In effect, I *intend* that you *believe* that we *understand* that the spirits *intend* that we act *with righteous intent.* At this point, we are at fifth order. Now we have what I would identify as a communal form of religion: there is something that we both have to agree on, and that something involves minds in another universe that can enforce their will on us.

I raise these two examples here specifically because, in many ways, they represent the essence of what it is to be human: culture, and the way in which we humans build the complex world of the imagination in which we live and breathe. It is my contention that nothing that we do (or any of our ancestors have ever done) comes close to being as cognitively demanding as story-telling and religion. Indeed, I have explicitly argued (Dunbar 2004) that our fifth order capacities really evolved to make religion possible. This is not for reasons intrinsic to religion itself (which should properly be seen as just another form of story-telling), but rather because religion has provided us with a mechanism for bonding (and bonding in a particularly deep sense) social groups that, by primate standards, are relatively large.

Conclusions

The burden of my case here has been that the primate and human brain (and, hence, mind) evolved to handle the intricate and dynamic complexities of a social world. This, naturally, was built upon a mammalian brain that was designed to deal with more everyday instrumental matters of physical cause and effect. But, in the case of primates, these fundamental and important cognitive processes have been refocussed (and, perhaps, enhanced) to deal with the less directly observable world of the mind that underlies social behaviour. While I would not wish to suggest that all primates have mindreading (or mentalising) abilities similar to those found among human children, I would want to suggest that they have a special sensitivity to social behaviour and the mental world. I have argued that this need to read and understand the mental worlds of others is computationally very demanding, and may be the principal explanation for why humans have enlarged neocortices (and especially frontal lobes).

If this case is accepted, then we need to ask fundamental questions about the relationship between the tasks that the mind evolved to do and the way we conventionally test for intelligence. The more benign answer to this question is that conventional instrumental intelligence (perhaps especially as reflected in *g*) underpins our remarkable abilities in the social domain. What is unique about human (and primate) social cognition is not the cognitive machinery as such, but rather the sensitivity to social cues. Hence, indices of intelligence such as those represented by conventional IQ tests correlate with performance in the social domain. However, it may be that social cognition involves unique features of its own that are not caught by conventional intelligence tests. If this is so, then it may be that conventional tests underestimate individuals' real abilities.

In one sense, this really raises the question as to what we think we are measuring when we apply intelligence tests. It cannot be doubted that they measure something of fundamental psychological importance. But we might pause to ask whether a crucial domain is missing from conventional instruments. This may be of practical significance given that we live in an intensely social world, and social skills are often as important in real everyday contexts as the ability to read, write and think logically. One reason for worrying about this is that some at least have claimed that humans (and, more generally, primates) may be able to solve problems when these are couched in terms of social contexts which they have difficult solving correctly when given as abstract or physical world cases (see, for example, Cheney & Seyfarth 1980, Tooby & Cosmides 1990).

Let me add one final plea. The study of individual differences as a topic in psychology has tended to focus exclusively on a combination of personality dimensions and intelligence in its conventional sense. I want to argue that this rather narrow remit should be broadened to include social cognition. Our studies have shown that there are striking differences between individuals in their capacities to engage in social cognition (e.g. advanced theory of mind), and I have suggested that these differences may not only have implications for social interactions but may also have consequences for individuals' abilities to perform in areas that are directly related to the nature of what it is to be human (namely, story telling and the creative arts). Failure to recognise this seems to me to leave the study of individual differences stuck in a vortex whose end point can only be a series of ever-smaller circles that eventually vanish. The lesson of what happened to learning theory – until the 1970s, one of *the* foundation stones of psychology – is a stark warning.

It is perhaps worth pointing out, in conclusion, that most of the findings reported in this chapter would not have been discovered but for the fact that we have approached human cognition from a comparative perspective. By being able to contrast humans against other animals (mostly primates, in this case), we have been able to highlight those features that are common to our animal heritage and those which seem to be unique. This use of animal comparisons has, I would argue, been crucial in allowing us to develop the kind of understanding of human social cognition that we have achieved in the past two decades.

References

Aiello, L.C. and Wheeler, P. (1995). The expensive tissue hypothesis: the brain and the digestive system in human evolution. *Curr. Anthrop.* 36: 199-221.

Astington, J.W. (1993). *The Child's Discovery of the Mind.* Cambridge (MA): Cambridge University Press.

Barrett, L., Henzi, S.P. & Dunbar, R.I.M. (2003). Primate cognition: from 'what now?' to 'what if?' *Trends Cogn. Sci* 7: 494-497.

Barton, R.A. (1996). Neocortex and behavioural ecology in primates. *Proc. R. Soc. Lond.* **263**B, 173-177.

Barton, R.A. and Dunbar, R.I.M. (1997). Evolution of the social brain. In: A. Whiten and R.W. Byrne (eds) *Machiavellian Intelligence II*, pp. 240-263. Cambridge: Cambridge University Press.

Byrne, R.W.& Corp, N. (2004). Neocortex size predicts deception rate in primates. *Proc. R. Soc. Lond.*

Byrne, R.W. & Whiten, A. (1988). *Machiavellian Intelligence.* Oxford: Oxford University Press.

Cheney, D.L. & Seyfarth, R.M. 91980). *How Monkeys See the World.* Chicago University Press: Chicago.

Clutton-Brock, T.H. & Harvey, P.H. (1980). Primates, brains and ecology. *J.Zool. Lond.*, 190: 309-323.

Dunbar, R.I.M. (1988). *Primate Social Systems.* Chapman & Hall: London.

Dunbar, R.I.M. (1992). Neocortex size as a constraint on group size in primates. *J. human Evol.* 22: 469-493.

Dunbar, R.I.M. (1993). Coevolution of neocortex size, group size and language in humans. *Behav. Brain Sci.* 16: 681-735.

Dunbar, R.I.M. (1998). The social brain hypothesis. *Evol. Anthrop.* 6: 178-190.

Dunbar, R.I.M. (2003a). Why are apes so smart? In: P.Kappeler & M.Pereira (eds) *Primate Life Histories and Socioecology*, pp. 285-298. Chicago: Chicago University Press.

Dunbar, R.I.M. (2003b). The social brain: mind, language and society in evolutionary perspective. *Ann. Rev. Anthrop.* 32: 163-181.

Dunbar, R.I.M. (2004). *The Human Story.* Faber & Faber: London.

Dunbar, R.I.M. & Bever, J. (1998). Neocortex size predicts group size in carnivores and some insectivores. *Ethology* 104: 695-708.

Dunbar, R.I.M., Duncan, N. & Nettle, D. (1995). Size and structure of freely forming conversational groups. *Human Nature* 6: 67-78.

Dunbar, R.I.M., McAdam, M. & O'Connell, S. (2005). Mental rehearsal in great apes and humans. *Behav. Processes* 69: 323-330.

Finlay, B.L. & Darlington, R.B. (1995). Linked regularities in the development and evolution of mammalian brains *Science* 268: 1578-1584.

Frith, C. (1996). Brain mechanisms for having a "theory of mind". *J. Psychopharm.* 10: 9-16.

Gibson, K.R. (1990). Cognition, brain size and the extraction of embedded food resources. In: J.G.Else & P.C.Lee (eds) *Primate Ontogeny, Cognition and Social Behaviour*, pp. 93-103. Cambridge: Cambridge University Press.

Henzi, S.P., de Sousa Pereira, L.F., Barrett, L., Hawker-Bond, D., Stiller, J. & Dunbar, R.I.M. (submitted). Social cognition and developmental trends in the size of conversational cliques.

Jerison, H.J. (1973). *Evolution of the Brain and Intelligence.* New York: Academic
Press.

Joffe, T. H. (1997). Social pressures have selected for an extended juvenile period in primates. *J. Human Evol.* 32: 593-605.
Kinderman, P., Dunbar, R.I.M. & Bentall, R.P. (1998). Theory-of-mind deficits and causal attributions. *Brit. J. Psych.* 89: 191-204.
Kudo, H. & Dunbar, R.I.M. (2001). Neocortex size and social network size in primates. *Anim. Behav.* 62: 711-722.
Lewis, K. (2001). A comparative study of primate play behaviour: implications for the study of cognition. *Folia Primat.* 71: 417-421.
Morino, L. 1998. What dolphins can tell us about primate evolution. *Evol. Anthropol.* 5: 81-86.
Martin, R.D. (1981). Relative brain size and metabolic rate in terrestrial vertebrates. *Nature* 293: 57-60.
Mitchell, P. (1997). *Introduction to Theory of Mind.* London: Arnold.
Pawłowski, B.P., Lowen, C.B. & Dunbar, R.I.M. (1998). Neocortex size, social skills and mating success in primates. *Behaviour* 135: 357-368.
Pérez-Barbería, F.J.& Gordon, I.J. (2005). Gregariousness increases brain size in ungulates. *Oecologia* 00: 000-000.
Pérez-Barbería, F.J., Shultz, S., & Dunbar, R.I.M. (submitted). *Evidence for intense coevolution of sociality and brain size in three orders of mammals.*
Pinker, S. (2002). *The Blank Slate.* Allen Lane: London.
Reader, S.M. & Laland, K. (2002). Social intelligence, innovation and advanced brain size in primates. *Proc. Natl. Acad. Sci., USA,* 99: 4436-4441.
Roth & Leslie, A.M. (1998). Solving belief problems: toward a task analysis. *Cognition* 66:1-31
Shultz & Dunbar (2005). Both social and ecological factors predict ungulate brain size. *Proc. R. Soc. Lond.* 000B: 000-000.
Stiller, J. & Dunbar, R.I.M. (submitted). Perspective-taking and social network size in humans.
Stiller, J., Nettle, D., & Dunbar, R.I.M. (2004). The small world of Shakespeare's plays. *Human Nature* 14: 397-408.
Tomasello, M. (2001). *The Cultural Origins of Human Cognition.* Harvard university Press: Cambridge (MA).
Tooby, J. & Cosmides, L. (1990). The psychological foundations of culture. In: Barkow, J.H., Cosmides, L. & Tooby, J. (eds) *The Adapted Mind*, pp. 19-136. Oxford University Press: Oxford.
Tulving, E. (2002). Episodic memory: from mind to brain. *Ann. Rev. Psych.* 53: 1-25.

IQ and The Chinese Experience: A Treasury of Facts[1]

Louise Higgins

Introduction

Today I propose to give you some examples of IQ and the Chinese experience. In 1869, Sir Francis Galton wrote "that if the question of hereditary genius were thoroughly gone into....China would be found to be a perfect treasury of facts bearing upon it."(p.391). In this paper I hope to bring you some interesting facts from this treasury. I shall start with the historical development of psychology in China, because the Chinese context is essential to understanding today's position. Then I shall present a brief review of over 300 papers published in Chinese from 1986 to 2005 looking at test development and use; factors linked with IQ such as urban and rural differences; gender differences; family factors ; the single child policy; ethnic minorities; key schools and ordinary schools; intelligence and academic achievement; learning disabilities; and gifted children.

Before I tell you about the Chinese experience of I.Q., I must give you some background to the development of psychology in China. The Chinese invented so many things; silk, porcelain, chess, the magnetic compass, blast furnaces, gunpowder, clocks and they also claim to have produced some of the first psychological ideas. Confucius, for example, classified people by their intelligence and ability to learn into high, middle and low levels (Cai, 1994; Howard Smith, 1973; Yan, 1996).

Over 3000 years ago, the Chinese used a variety of methods of measuring talent and behaviour such as identifying intelligence by

[1] The help of Gao Xiang and the support of The Sino-British Fellowship Trust is gratefully acknowledged.

response speed and measuring mental attributes by interviews, in order to enable the Emperor to assess his officials' fitness for office, the 'ke ju' Imperial exams. The person who came first in these exams was known as the zhuang yuan and Galton was interested to note the case of a woman whose son was a zhuang yuan and who then remarried and had a son by her second husband who also became a zhuang yuan.

In Confucius' private school the six classical arts were taught. These six were:

1: Etiquette (rules, rites)	interpersonal intelligence
2: Music	musical intelligence
3: Archery	kinaesthetic intelligence
4: Arithmetic	logical – mathematical intelligence
5: Writing	verbal intelligence
6: Riding	spatial intelligence

These could be seen as similar to Gardner's theory of multi-intelligence (Gardner, Kornhaber and Wake, 1996).

Intelligent people were those whom ancient philosophers (see Yan, 1996) thought could:

1) infer and predict from given information,

2) know and understand self and other people,

3) find relationships between things, also between people and things,

4) be creative and deal well with the unexpected,

5) make full use of other people's wisdom and intelligence.

The last of these, collective wisdom, would be the highest form of intelligence. It derives from the Chinese collective culture and is different from the way we see things in the West.

By the time of the Han dynasty (BC206 – 220AD) the use of multiple test batteries in the Chinese civil service examination was common, with essay writing and oral examinations on, for example, civil law, agriculture, taxation and geography. By the Ming Dynasty

(1368–1644AD) complex national and multi-stage testing programmes had developed.

Another example of testing from ancient Chinese folk culture about 6BC was "testing the child at one year of age". On a child's first birthday, he/she would be placed on a large table covered with items of food, clothing, paper, toys, books, bows and arrows for boys and needle and thread for girls. The baby was encouraged to crawl about and pick up the items he/she liked best. By observing what the baby grasped first, the parents predicted the child's aptitudes and personality. Clearly this is not a test by modern standards, but it does show willingness to assess individual differences by concrete means.

Liu Xie (a 6th Century scholar) asked people to draw a square with one hand and a circle with the other to show that people cannot do two or more things simultaneously because of interference. Binet, who developed the famous modern intelligence test, used a similar task in the 1890s to show the effect of distraction on mental tests.

Tangrams, a kind of jigsaw puzzle, were invented during the Song dynasty and could be seen as the first test of creativity, while the puzzle ring was the earliest problem solving test.

Sadly although there were all these early psychological ideas, they were never organised systematically or scientifically in China. However, the British East India Company decided to adopt the Chinese system of civil service exams to select employees from 1832 and later the British Government saw that this worked well and they adopted the idea in 1855, so perhaps we could say that the Chinese did invent mental testing.

In the late 1880s new Western ideas began to reach China. Chinese students started to study psychology in the West and translate Western psychology books (the first in 1889). From the 1920s to the 1940s, Chinese psychology was essentially Western and experimental. But with the founding of the People's Republic of China, in 1949, there was a move to reform the social sciences under the guiding principles of Marxism and only experimental psychology (particularly the Soviet Russian model) was seen as appropriate. The psychology of individual differences was anathema to Marxists who saw it as a 'tool of the bourgeoisie' which contradicted the Marxist doctrine that people are shaped by their social class. Later, I shall present research evidence which suggests that social class is still seen as the major influence on individual differences by Chinese psychologists today.

Between 1966 and 1976 (The Cultural Revolution) psychology as a discipline was attacked as a 'bourgeois pseudo-science' and thrown out as a scientific discipline. University psychology departments were closed down and professors sent out to the countryside to labour on farms and down mines. This has been described as a dark period for psychology in China, and a great price to pay for political interference in science.

Fortunately, with the Open Door policy of the late 1970s and early 1980s, psychology was rehabilitated and it is now re-established and increasingly supported by the Chinese Government, especially in the applied field where, with the Socialist Market Economy policy, it is now recognised that economic progress ultimately depends on the talents of managers and workers. The Government now sees the value of investing in modern psychology. However, psychology is still a small discipline, with fewer than 2 psychologists for every million people in China, compared with about 180 per million in the UK.

The development of I.Q. tests

Pre-Liberation, the Binet Scale, was introduced in 1916. It was translated and revised several times. The Goodenough Draw a Person test and a few others were also translated and used in Southern China.

Liberation was in 1949. There was then a gap till the late 70s (post Cultural Revolution) when revisions of the WAIS (Wechsler adult intelligence scale), WISC (Wechsler intelligence scale for children), and the Wechsler Preschool tests were completed by Professor Gong Yao Xian who, with his team, translated or developed more than half the tests used in China today. A lot of work has been done on the reliability and validity of these tests, and on obtaining large and representative samples based on censuses.

It was not just a case of translating the language, but also of relating the questions to the Chinese culture, for example references to Alice in Wonderland were changed to A Dream of Red Mansions, references to Christmas changed to the Chinese Spring Festival and references to hobbies like golf were changed. Gong always gave both urban and rural norms for his Chinese samples and even had two slightly different test forms, for instance, with questions about farm life for rural children.

Cheng and Zheng (2000) published a review of I.Q.testing using the WAIS – RC(1981-1998) taken from106 publications. The following purposes were noted:

39 neurological assessments

34 intelligence assessments

12 effect of treatment studies

9 criterion matching for other scales

8 diagnosis

4 others

The tests were mostly used in clinical settings, then for research, then by expert court witnesses, then for educational purposes.

In China, teachers are not regarded as qualified to use I.Q. tests but doctors and psychologists are. Most schools in cities now have a psychology graduate on the staff who does counselling. Teachers can refer pupils to them or children can be referred to the local hospital with a psychology department. Individual testing is not done a lot because the tests are expensive. Children's mental health records often include personality test scores, and Raven's Progressive Matrices IQ scores measured every three years to check progress (The Ministry of Education of the People's Republic of China, 2002). This Western test has been revised several times and Chinese norms produced. Raven's Progressive

Matrices is important because it is a non-verbal test, and is, therefore, especially useful for testing minority nationalities who do not speak Mandarin.

Increasingly the Chinese are developing their own scales, for instance Professor Gong's Chinese Intelligence Scale for Young Children, (Dai, Gong & Tang , 1998; Tang, Gong & Dai, 1998;). Another test is Gong's Test of Nonverbal Intelligence. (Gong, 1999; Gong, Zhang & Cheng, 1995; Ren & Gong, 2002). This test of non verbal intelligence was originally devised for testing ethnic minorities but now there are national norms available. A good English language summary of all these test developments is available in Shi (1995).

Urban Rural Differences

There are consistent findings of significantly higher I.Q. in city children compared with rural children (Zhang, Liu and Zhao, 1998; Zhu and Fu, 2002). This difference was greatest at age 5 and then declined and was explained as due to the city children's access to kindergarten which gave them an early advantage. The lower I.Q. of country children is generally attributed to socioeconomic factors such as poverty, having better teachers in schools in the cities, the education level of parents in cities being higher, and early educational provision being better in the cities. Even within rural areas there are significant differences in levels of deprivation, for example, IQ scores were lower in rural children in Jiangxi Province (87.54±11.54), which is remote and mountainous, than in Zhejiang Province(91.42±13.67), which is richer and by the sea.

Gender Differences

Many studies (Guan, Fu and Cui et al., 1997; Luo, Guan and Li, 2001; Peng and He, 1998; Qin, Wang and Hou, 2004) showed that boys had higher I.Q. scores but these differences were not significant. However Guo and Lu (1999) found significant gender differences in the countryside, with boys scoring higher than girls.

N=501		Mean IQ	SD
M=242	combined Raven's	107	15.9
F=259		102	15.6
		T=2.8648 $p<0.01$	

It is suggested that this result is caused by parents paying much more attention to boys in the countryside, whereas in the cities parents treat boys and girls more equally. It was also found that girls in the cities appeared to be ahead of boys at the age of 9 to 10 and this was attributed to girls' being ahead in physical maturation. Physical maturation comes later for boys and girls in the countryside in China. West, Zhao, Chang and Cheng (1999) provide a good background to the role of women in China today.

Single Child Policy

Before this policy became law in 1979/1980 there were not many singletons. It was initially implemented more rigorously in the cities than in the countryside. Ethnic minorities may have more than one child.

Single children are consistently found to have higher I.Q. than those with siblings (Luo, 2004; Wang, Xu and Li, 1997). With only one child, parents try to give him or her every opportunity that they can and this maximises intelligence development.

Actually, although there is a lot of research about single children, it is mainly about their personality, mental health and behaviour problems. Results suggest that being a single child is not necessarily a good thing. More research is ongoing in this area.

Key schools and ordinary schools

In the 1980s, there were selection examinations, which involved language, mathematics, and English, at age 11 years and again after junior middle school (14 years) and the top scoring children were then sent to key schools. Now selection is only at 14 as educational reforms were implemented to "lighten the load" of primary school children. However, in Chinese schools, most knowledge still has to be memorised from textbooks.

In key schools the children's average I.Q. was found to be 124.3, with 40.5% above 130 and 81% above 120.(Wang, Xu and Li, 1997). It was noted that I.Q. differences between children in key schools and in ordinary schools decreased with age. This was put down to the later intellectual maturation and catching up of children in ordinary schools.

The best teachers are sent to key schools where children have to study hard and do a lot of homework. They have no spare time for their own interests. However in ordinary schools pupils have more time for their own development, and to learn what they want. This helps their intellectual development. There are new educational reforms planned to widen the extremely 'exam based' education in schools. Becker (2000) and Starr(1998) provide an interesting background to education in China.

Age Differences

There is some research which suggests that within the same grade, the I.Q. levels of younger children are higher. Everyone starts school at the age of 6 but some may start school early, and these children often have higher IQ and may progress through the whole education system early (Wang, Xue and Li, 1997; Yan, 1997).

I.Q. and academic achievement

Many Chinese studies (Lin, 2002; Shen, Li and Zhuang, et,al. 2000; Wang and Duan, 1999; Xiao, Guan and Tang,et,al. 1999) have demonstrated the strong link between high I.Q. and good academic achievement. However, Shao (1990) found decreasing correlations between I.Q. and achievement from primary school, 0.6 – 0.7, to middle school, 0.5 – 0.6, and then to college, 0.4-0.5. Why should this be? It is possibly because of teaching methods. In primary school those who study hard and memorise will do well, whereas in Universities they learn how to learn. University entrance exams in China are changing and will in future test how to apply knowledge more than rote learning. Also, in primary schools, more modern teaching methods are being developed.

IQ has also been found to predict success at university entrance exam level, but it was not the only factor. High IQ was necessary but not sufficient as those who got the top marks were high IQ scorers but not the highest.

Family Factors

Many studies concentrate on parents' careers, parents' education and children's early education (Guan, Fu and Cui,et,al. 1997; Luo, 2004; Luo, Guan and Li, 2001; Qin, Wang and Hou, 2004; Wang, 1997; Wang, Xue and Li, 1997; Wei, Dou and Song, et,al.2000; Yuan, Chen and Li, 1998; Zhang, Guo and Li, 1998). Most suggest that mother's education is more important for children's IQ than fathers' education, both in the countryside and in the cities. In the West this is explained as being due to the 'maternal environment'.

There is found to be a positive relationship between higher socioeconomic class and children's IQ, because richer parents are able to pay for kindergarten, music lessons, travel, and so on, whereas, parental problems such as illness, and addictions are linked to lower children's IQ. The parents' actual career is also linked to the IQ of the children as below:

	Father	Mother
Highest	Teacher	Teacher
	Scientist	Government
	Government Worker	Scientist
	Worker	Worker
Lowest	Self Employed	Self Employed

Minority Nationalities

There are 56 races in China including the Han majority. There are 106,643,000 minority people. Some background to the position of Chinese minorities may be found in Gittings, 1996 and Tu, 1994. Most of the minorities live in rural and Western China. Minority IQ studies concentrate on: 1) Comparing minorities with the Han, 2) Comparing minorities who live together with those who live with the Han and 3) Bilingual education. These studies largely use non-verbal tests because of language differences. The

IQ of minorities is consistently found to be lower than that of the Han. Chinese psychologists say we should be cautious in explaining these results and drawing conclusions.

1) Comparing minorities with the Han.

Na and O, (1994) reported the following:

<table>
<tr><td colspan="2">Inner Mongolians</td><td>N=1481</td><td colspan="3" rowspan="2">aged 5-10 yrs in Xinjiang Province</td></tr>
<tr><td>Han</td><td></td><td>N=3213</td></tr>
<tr><td rowspan="2">Raven's scores</td><td>Mongolians</td><td>96.32</td><td>+/-</td><td>11.08</td><td></td></tr>
<tr><td>Han</td><td>101.59</td><td>+/-</td><td>12.6</td><td>(p<0.001)</td></tr>
</table>

In 1990, a sample of Miao children in rural Guizhou, aged 3 -6 years, were tested. Their IQ scores were found to be lower than the Han. However, these researchers used Professor Gong's revision of the WISC which was all in Mandarin and, therefore, very unsuitable. They did also report significant gender differences in the Miao children, with boys scoring an average of 80.12 and girls 76.64 p<0.05 (Zhang, Tu and Zheng, 1990).

Like urban- rural differences these low scores by minorities are seen as due to social-economic class differences. It might not be acceptable to write about racial genetic differences, or about marriage with cousins in these groups. Researchers in the People's Republic of China are very concerned about the effects of environment and parenting on children's intelligence. However, there have been some twin studies which are the classic method of investigating heredity/environment influences (Bai, Duan & Chen, et, al. 1994; Li, Chen & Ma, et. al. 2001; OuYan, Wang & Wang, et. al. 1996; Xu, Zeng & Zhang, et. al. 1996; Zhen, Ji & Yan, et. al. 2002;). These have shown a positive correlation between the IQ, language and maths scores of twins, highest in monozygotic twins, next in dizygotic same sex twins, and then in dizygotic opposite sex twins. The reported correlations range from 0.42 to 0.91and the effects of heredity decrease with age, so the researchers conclude that heredity is involved but is not the whole story (Lin, 1981).

In 1994, the WISC was translated into the Mongolian language, but not revised for their culture, and given to children aged 5 to 6 years (Yang and Gong).

		Full IQ	Verbal IQ	Performance IQ
Mongolian	150	94.9	89.9	100.8
Han	151	99.1	100.1	98.3
		P<0.01	P<0.01	P<0.05

The sample was taken from Inner Mongolian areas where the Mongolians live together with the Han. A non verbal scale was also used to test both groups with the following results:

	N	Full IQ	Completing Figs	Rearranging Figs	Missing nos
Mongolian	60	101.3	10.6	10.5	10.1
Han	60	98.7	9.4	9.5	9.9

While the difference between the full IQ scores was not significant, all the other differences were significant at $p<0.05$ level, demonstrating that the Mongolian children did not always score lower than the Han.

2) Research comparing those who live in minority areas with those who live with the Han.

Those who live with the Han usually score higher. They tend to live near cities and speak Mandarin while those who live in their own communities only speak their own language and are poorer.

Samples of Zang, Dai, Yi, and Miao peoples aged 8 to15years were tested in 1997 by Cai and Ding and found to be lower than the Han. However, like the Han, their IQ was shown to increase with age, especially from 8 to 12 years. There were some significant differences between the minorities, for example, the Zang scored significantly higher scores on Raven's Matrices than the other minorities. The conclusion from these studies is that while inheritance determines potential, social environment determines the level which is actually reached. It was noted that those who lived with the Han had higher speeds in all tests at all ages than those who lived in their minority communities, indicating that the speed factor in testing may not be relevant in other cultures.

3) Bilingual Education

This has developed rapidly since 1977 especially in the Zang autonomous region (Tibet). Studies by Xing, 2001, and Yu, 2003, have investigated children who are either educated entirely in Mandarin, or in two forms of bilingual education, namely, mainly Zang with some Han, or mainly Han with some Zang. A longitudinal study on Grade 2 to Grade 7 children was carried out on 90 Zang children taking bilingual education and 29 who only learned Mandarin. Raven's Matrices was administered as a pre and post test. There were no differences between the groups in the pre tests but there was a significant difference in the post tests when the single language (Han) group obtained higher scores.

After 5 years, the boys in the bilingual group gained higher scores than the girls in the bilingual group, and the single language girls did better than bilingual girls. It was, therefore, concluded that bilingual education was not good for girls. The results were explained as an interaction between gender and type of language teaching, especially for the girls. For girls the single language teaching appeared to be the best, the Zang/Han next, and the Han/Zang the least good. Was bilingual education too difficult for girls? Was there not enough time to learn other things and why was it different for boys? More study is continuing.

Low IQ

In 2000 there were 10,170,000 people including 5,390,000 children (aged 0-14) in China with IQ less than 70, therefore, in view of the size of this problem, more and more research is being done.

Several studies investigated the problems of the learning disabled in school and found that the mental health of those in normal schools was poorer than those in special schools. (Li, 1998; Wang and Wang, 2004; Zhang, 2002; Zhang, Jiang and Yang, 2004; Zhao and Zuo, 2000). Two studies reported more parental rejection, overprotection and severe punishment among the learning disabled children (Jiang, 2004; Ma, Zhang and Jiang, 2005). Further research has been carried out on the behaviour problems of learning disabled children at school and their personalities, by Zhang (2002). Li (1998) showed that learning disabled children in rural areas have fewer problems, perhaps this is because country people are kinder to them. There have been recent successful attempts to improve learning disabled children's attention and concentration span by calligraphy therapy (Liu, 1999).

There has also been a lot of research on children with IQs over 70 but less than average, the lowest 10% of children in normal schools. It was concluded that IQ is the main factor affecting school record but not the only one. These children tended to have lower verbal IQs than performance IQs. They showed limited acquired knowledge, deficits in calculating ability, speed and accuracy but they also showed relatively good spatial and perceptual/organisational abilities. This imbalance between verbal IQ and performance IQ, very common in learning disabled children, together with their attention deficits, may help to explain their low achievement in school (Ding and Feng, 2002; Ji and Du, 2000; Ju, Chen and Zhang, 1999.)

Other environmental effects

A natural experiment on the effect of extreme deprivation on IQ development was reported by Mei (1991) from a very poor rural area in Hebei and Shandong Provinces where infants from the age of 10 days until 1½-2 years were placed every day in sandbags which acted as nappies and served to prop them up. They were fed but not talked to and generally left on their own for up to 16 hours per day. They did not cry and were very quiet. These children were tested using the rural form of Raven's Matrices and the Chinese Binet revision and not surprisingly were shown to have IQs on average 5% lower than normal in the region. 15.9% of them had IQs lower than 69.

		IQ	SD	N	P
Raven	Sandbag	89.07	19.42	261	<0.01
	Normal	93.86	15.82	220	<0.01
Binet	Sandbag	77.00	12.61	161	<0.01
	Normal	80.78	11.64	220	<0.01

Chen, Zhou and Bai (1998) noted that children in 'reform schools' had lower IQs than children in normal schools. It was also noted that 10.2% came from single parent families (broken by divorce and death) and 44% had had poor parenting. There are more and more single parents in China today. There is also more delinquency and more drug problems as the society faces the challenges of modernization.

A recent paper shows that China is adopting the American Association for Mental Retardation's revised definitions of retardation with more stress on functioning level and quality of life and less on IQ. It noted that support from school administration and teachers is not enough, suggesting that more school funds are needed and more information needs to be provided for teachers and parents (Xu, 2003).

Gifted children

There are few research papers but quite a lot of government reports and conference statements on this topic. Some schools provide special classes such as the Beijing 'Ace' middle school which has sent over 100 children to university aged 13–16. Several universities have special classes for young children or sometimes put them in mixed classes with the best students. Out of a sample of 500 graduates from these special classes, 70% went on to Masters degrees, and more than 100 obtained a PhD, the youngest at 23 years of age, and these qualifications were not just in the sciences. (http://www.gmw.cn (in Chinese))

In July 2005, a 10 year old boy was the youngest ever to pass the college entrance examination to do applied maths at university. However, people are worried about how well he will cope socially and emotionally at university. (http://www.21cbe.com (in Chinese))

There are thought to be 10,000,000 children in China with IQs over 130 and there is much national discussion about why China should have special educational facilities for these children. Educators say that normal schools may not meet the needs of highly intelligent children and since there are already special provisions for the deaf and the learning disabled it would be fair to offer gifted children special education. It is important for China to develop these children's full potential and surveys show that the majority of people support the idea of special provisions in science, liberal arts, music and sport. However, others note that early selection might miss late developers and that textbooks and teacher training are not good enough at present. There are also concerns about how gifted people will relate to others when they start work as it is important for education to produce well rounded people and not just maths geniuses. (http://www.cosn.net/htm/special (in Chinese))

Conclusion

Overall when Chinese researchers look at IQ they stress environmental influences. In many areas of research their results are the same as those found in the West despite coming from very different societies, supporting the value of cross cultural studies in demonstrating universal truths about human beings. There are also, in the Chinese researches, some extremes of deprivation, isolation and scale to which Western psychologists do not normally have access in their investigations.

Psychology is of growing importance in China today, and I am confident that the future will bring very many new and exciting Chinese research findings in the field of intelligence.

References

Becker, J. (2000). *The Chinese.* London: John Muray.

Cai, X. Q. (1994). *Analects of Confucius with English translation.* Beijing: Sinologue.

Cai, X. Y. & Ding, N.Y.(1997). The comparative study on 8-15 yeas old children who lived in the minority area and others lived with the Han. *Nationality Study*, **4**, 61-68. (in Chinese)

Chen, R., Zhou, L. & Bai, C. Y. (1998). A study on the intelligence structure and influence factors of students in reform school. *Chinese Journal of School Doctors*, **6**, 407-409. (in Chinese)

Cheng, Z. H. & Zheng H. (2000). The application of WAIS-RC in China. *Chinese Mental Health Journal*, **4**, 235. (in Chinese)

Ding, Y. H. & Feng, L.Y. (2002). Analysis of the intelligence level and structure of children with learning difficulties. *Chinese Journal of Clinical Psychology*, **3**, 227-228. (in Chinese)

Galton, F. (1869). *Hereditary Genius*.2nd Ed.1962, The Fontana Library, London, W.Collins & Sons.

Gardner, H., Konhaber, M. L., Wake, W. K. (1996). *Intelligence: multiple perspectives* Fort Worth: Harvard Brace College Publishers.

Gittings, J. (1996). *Real China: From cannibalism to karaoke*. London: Pocket Books.

Gong, Y. X. & Cai, T. S. (1994). WPPSI revised in China. *Chinese Journal of Clinical Psychology*, **1**,1-6. (in Chinese)

Gong, Y. X. (1999). Norms of the nonverbal intelligence tests on Chinese children of Han nationality. *Chinese Journal of Clinical Psychology*, **3**,129-134. (in Chinese)

Gong, Y. X., Zhang, Z. X. & Cheng, Z. H. et al. (1995). Norms of the non-verbal intelligence tests on the minority nationalitities. *Chinese Journal of Clinical Psychology*, **3**, 129-136. (in Chinese)

Guan, X. L. Fu, M. S. & Cui, Y. H. et, al. (1997). The development of children's intelligence and the influencing factors. *Journal of Shang Dong Medicine University* (Social Science Edition), **4**,14-16. (in Chinese)

Guo, X. Q. & Lv, J. Z. (1999). Study on primary school students' gender difference of intelligence. *Journal of HeNan Education College* (Social Science Edition), **1**,19-23. (in Chinese)

Howard Smith, D. (1973). *Confucius* St. Albans: Paladin.

http://www.21cbe.com (in Chinese)

http://www.blogchina.com/new/display/85129.html (in Chinese)

http://www.gmw.cn (in Chinese)

Ji, W. D. & Du, Y. S. (2000). Study on the intelligence structure of learning disabled children. *Health Psychology Journal*, **4**, 403-405. (in Chinese)

Jiang, Q. E. (2004). A study on the relationship between parenting styles and mental health of mild mental retardation children. *Chinese Journal of Special Education*, **7**, 38-41. (in Chinese)

Ju.W., Chen, L. P. & Zhang, Z. H. et. al., (1999). Study on the intelligence structure of different kinds of learning disabled children. *China School Health*, **6**, 441. (in Chinese)

Kruger, R. (2003). *All under heaven: a complete history of China*. Chichester: John Wiley & Sons Ltd.

Li, Z. S. (1998). Research on the personality characteristics of retardation children. *Journal of ChongQi Normal College* (Natural Science Edition), **6**, 74-80. (in Chinese)

Lin, C. D. (1981). The effect of heredity and environment on children's intelligence development: psychological research on twins. *Journal of Beijing Normal University* (Social Science Edition), **1**, 64-73. (in Chinese)

Lin, H. (2002). The relationship among intelligence, family environments and academic achievement. *China School Health*, **2**, 167-168. (in Chinese)

Liu, Y. (1999). Attention recovery effect of calligraphy training for MR children. *Chinese Journal of Special Education*, **3**, 39-41. (in Chinese)

Liu, Y. S. (1998). Factors in examinations results in college entrance examination. Journal of JiAn Normal College (Natural Science Edition), 5, 46-51. (in Chinese)

Luo, W. Q. (2004) The influencing factors analysis on 600 students in grade one of primary school. *Chinese Journal of School Doctors*, **2**, 139-140. (in Chinese)

Luo, Z. W., Guan W. & Li, Y.(2001). Study on pre-school students' intelligence and the influencing factors. *Chinese Primary Health Care*, **12**, 60. (in Chinese)

Ma, Z. Z, Zhang, F. J. & Jiang, L.(2005). The regression analysis of the mental health of parents and their mentally retarded child's behaviour problems. *Chinese Journal of Special Education*, **4**, 46-49. (in Chinese)

Mei, J. (1991). A study of the IQ of sandbag-raised children. *Psychological Science in China*, **1**, 42-44. (in Chinese)

Ministry of Education P. R. C. (2002). The direction outline of mental health education in primary and middle school. Retrieved September 5, 2003 from http://www.moe.edu.cn/news/2002_9/10.htm (in Chinese).

Na, D., O, E.L. (1994). The results of 5-14 Mongolian children's Raven Tests in XinJiang. *Journal of NeiMeng Medicine*, **1**, 37-38. (in Chinese)

Peng, C. Q. & He, Q. D. (1998).The results of analysis on 420 primary school children's intelligence tests. *Journal of HengYan Medicine College*, **1**, 72-73. (in Chinese)

Qin, R., Wang, Z. X. & Hou, H.Y. (2004). The development of intelligence of primary school students in BaoTou and the influencing factors. *China School Health*, **4**, 460. (in Chinese)

Shao, R. Z. (1990). *Psychology of teaching and learning*, China East Normal University Press. (in Chinese)

Shen, D. L., Li, H.Y.& Zhuang, S. F. et,al. (2000). The correlation analysis on the intelligence, study attitude and the math achievement in secondary and primary school. *Journal of TianJin Normal University* (Elementary Education Edition), **2**, 1-5. (in Chinese)

Shi, J. N. (2004). Diligence makes people smart: Chinese perspectives of intelligence. In R. J. Sternberg (Ed.), *International Handbook of Intelligence* (pp.325-343), Cambridge: Cambridge University Press.

Starr, J. B. (1998). *Understanding China*. London: Profile Books Ltd.

Tu, W. M. (1994). *The living tree: The changing meaning of being Chinese today.* Stanford, California: Stanford University Press.

Wang, L. Y. (1997). The effect of rural family environmental factors on children's intelligence development. *Journal of JiangShu Predictive Medicine*, **4**, 49-50. (in Chinese)

Wang, W. H. & Duan, J. Y. (1999). The correlations among intelligence, creativity, creativity trend and academic achievement. *Journal of XiYan Normal College* (Social Science Edition), **1**, 70-72. (in Chinese)

Wang, W. N., Xue P. & Li, B. (1997). The study on key school students' intelligence level and the correlated factors. *Primary Education*, **4**, 31-32. (in Chinese)

Wang, Y. & Wang, J. Y. (2004). Research on the school adaptive behaviours of students with mental retardation. *Chinese Journal of Special Education*, **6**, 30-34. (in Chinese)

Wei, X., Dou, G. & Song, Z et, al., (2000). The relationships among parent's careers, educational levels and children's intelligence. *Journal of YunNan Normal University*, **5**, 18-24. (in Chinese)

West, J., Zhao, M. H., Chang, X. Q., Cheng, Y. (1999). *Women of China: Economic and Social Transformation.* London: Macmillan Press Ltd.

Xiao, X. Q., Guan N. H. & Tang, J. X et, al., (1999). The correlation analysis on the intelligence and academic achievement in rural primary school. *Health Psychology Journal*, **1**, 21-23. (in Chinese)

Xing, Q. (2001). On the intelligence state of Tibetan bilingual children in difference community culture and bilingual model. *Journal of South China Normal University* (Natural Science Edition), **3**, 49-54. (in Chinese)

Xu, J. C. (2003). On the evolvement of the new definition of mental retardation: the new tendency oriented by function, support and quality of life. *Chinese Journal of Special Education*, **4**, 19-23. (in Chinese)

Yan, L. S. (1996). The evaluation criterion of intelligence in ancient China. Journal of HuNan Normal University(Social Science Version), 3,46-50.

Yan, G. F. (1997). The development level of intelligence of pre-school students and the correlated factors. *China School Health*, **5**, 344-346. (in Chinese)

Yang, W. P. & Gong, Y. X. (1994). The comparative study on 5-6 children's intelligence between the Mongolian and the Han. *Chinese Journal of Clinical Psychology*, **2**, 81-85. (in Chinese)

Yu, Q. (2003). A comparative study of intellectual development of boys and girls in bilingual education. *Journal of Research on Education for Ethnic Minorities*, **3**, 44-49. (in Chinese)

Yuan, X. Q., Chen, X. X & Li, C. (1998). Intelligence levels and influencing factors of 7-13 years old children in HengYan City. *Applied Preventive Medicine*, **2**, 72-74. (in Chinese)

Zhang, J. Q., Liu, Y. X. & Zhao, L. S. et, al. (1998). The research on the intelligence of primary school students living in the city and those living in the rural area. *AnHui Preventive Medicine Journal*, **2**, 168-169. (in Chinese)

Zhang, Z. Y., Guo, X. W. & Li, S. L. (1998). The influencing factors on children's intelligence development in the urban areas, *China School Health*, **6**, 439-440 (in Chinese)

Zhang, D. R., Tu, L. H. & Zheng, Y.(1990). The comparative study on the Miao and the Han children's intelligence in GuiZhou. *Chinese Journal of Nervous and Mental Diseases,* **1**, 9-11. (in Chinese)

Zhang, F. J. (2002). A research on the adaptive behaviour development of children with mental retardation. *Psychological Science*, **2**, 170-254. (in Chinese)

Zhang, F. J., Jiang Q. E & Yang, F. Y. (2004). A mental health research on children with mild mental retardation. *Psychological Science*, **4**, 824-827. (in Chinese)

Zhao, Q. M. & Zuo, Y. F. (2000) Research on mild mental retardation students' intelligence, self-control and adaptive behaviour disorder. *Journal of Luliang Higher College*, **4**, 34-37. (in Chinese)

Zhu, Q. Y., Fu, G. Y. (2002). A comparative analysis by using Draw A Person Test. *Chinese Journal of Clinical Psychology*, **1**, 54-55. (in Chinese)

Follow-up Studies of the Scottish Mental Surveys of 1932 and 1947

Ian J. Deary

The Scottish Mental Surveys of 1932 and 1947

In Scotland, there are data on the IQs of almost everyone born in 1921 or 1936 and who was at school in Scotland at age 11. This essay describes some of the research that has been conducted using these data.

The Scottish Mental Surveys of 1932 (SMS1932) and 1947 (SMS1947) might be the only examples of near-entire populations being IQ-tested. On June 1st 1932, for the avowed purposes of finding the incidence of mental deficiency and of planning for educational provision, the SMS1932 tested almost all Scottish schoolchildren born in 1921. The massive exercise was executed by the Scottish Council for Research in Education (SCRE, 1933). The SMS1932 was in part encouraged by SCRE's involvement in the International Examination Inquiry, and some funding was provided by the Carnegie Corporation of America. On June 4th 1947, partly because of worries that "intelligent parents are limiting their families" (SCRE, 1949, p. vii), the mental exercise was repeated, on almost all Scottish schoolchildren born in 1936. One aim was to discover whether ability test scores had fallen between 1932 and 1947. They hadn't: the SMS1947 had a higher average score than the SMS1932. The chairman of the SMS1947 was Godfrey Thomson, Professor of Education at the University of Edinburgh. Sponsorship of £4,000 for conducting the SMS1947 came from the Population Investigation Committee, and funding of £2,000 for publishing the results came from the Eugenics Society (now the Galton Institute). Therefore, the present chapter might be viewed in part as a slightly late report on a grant that was awarded about 60 years ago.

Both the SMS1932 and the SMS1947 used the same mental test, which was a version of Thomson's Moray House Test No. 12.

The Moray House Test series were used for primary to secondary school selection in England. Although the test had the disadvantage that it was thought to be rather difficult, it had the advantage that it could be used for no cost. It tends to be referred to as the 'Verbal Test' in SCRE's reports, but there are some numerical, spatial and other types of items, though verbal material predominates. To ensure that measurement was adequate at the below-average ability range, there were additional non-verbal tests.

Apart from the basic data provided by the whole-population survey based on the Moray House Test, there were some additional collections of data in both surveys.

In the SMS1932, 500 boys and 500 girls were tested individually on the Stanford-Binet test. This provided concurrent validity for the Moray House Test. The two tests correlated at about 0.8 (SCRE, 1933).

The additional data collected from some of the SMS1947's participants was more complicated. Additional sociological data were collected from the persons who were born on the first three days of each month (around 7000 people). These were referred to as the 36-Day Sample. Similar data were collected from each pair of twins in the sample. Even more detailed information was collected on the persons born on the first day of the even-numbered months (around 1200 people). This group is known as the 6-Day sample. They were administered the Terman-Merrill revision of the Binet test and, thereafter, interviewed at home for the next sixteen years (Maxwell, 1969).

With some minor losses over time, these were the data that were held by the Scottish Council for Research in Education when Professor Lawrence Whalley and I became aware of these surveys in 1997. Information on two whole populations of older people's IQs at age 11 appeared to be a potentially useful tool for research. To date, there are about 50 recent peer-reviewed publications that have included these data, so it will be possible to offer only some examples of the research output from the Mental Surveys' data. The research tends to be of three types. First, there are 'psychometric studies.' These are research reports that use the stored data, without involving further data collection. Second, there are 'cognitive epidemiology' studies. These studies involve our linking the data in the Mental Surveys to other databases, such as records of deaths and illnesses. Third, there are 'cohort studies.' These involve recalling people who took part in the Mental Surveys and re-testing them.

Psychometric studies using the stored data

Because the SMS1932 contained essentially an entire population's mental test scores, it was possible to inquire about sex differences almost without sampling bias (Deary, Thorpe et al., 2003). The data from the Moray House Test and the non-verbal tests were combined to give a single score, and then boys' and girls' scores were compared. Despite there being around 40,000 boys and girls, offering large power to detect even a small difference, there was no difference in means between the two sexes. Therefore, at age 11 years, on a general mental ability test, there was no mean difference between boys and girls. However, they did differ significantly in the spread of their scores. The standard deviation in the boys was greater than that in the girls. The effect of this was that there were more boys with extreme scores, both high and low. For example, at IQs around 65 and 130+ there were about 56% boys and 44% girls. This increased variance in boys might in part account for their being found in greater numbers among people with mild mental handicaps, and among people with high achievements such as first class university degrees.

Some data that were republished from the whole SMS1947 survey was the association between IQ and family size (Deary, Whiteman, Starr et al., 2004). This showed that participants in families with just one or two children had the highest Moray House Test raw scores (above 40 points), and that there was a near-linear decrease in scores with greater sibships, such that participants in families where there were more than 13 children had scores below 25, around a whole standard deviation lower. Figure 1 shows the same data, for sibships up to seven, in a new form. The Moray House Test scores are plotted for position in family and as a function of family size. Family size has the major effect. There is some tendency for the first born and the last born children to have higher ability test scores. The original publication (Scottish Council for Research in Education, 1949) discussed various caveats from these data; for example, the older children from large families might be misclassed because their families were still incomplete.

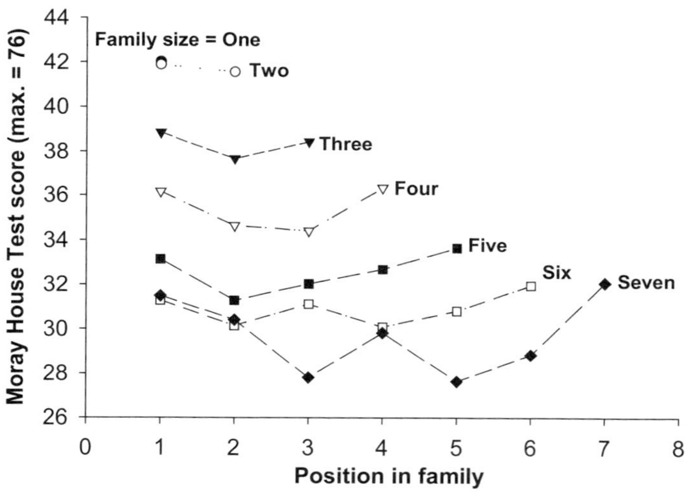

Figure 1: Moray House Test score by family size and position in family in the Scottish Mental Survey of 1947 (computed from Scottish Council for Research in Education, 1949, p. 107).

The explicit identification of the twins in the SMS1947 offered an unusual opportunity to address an entire population of twins. Twin studies typically examine samples of a population's twins, with the possibility of bias in the sampling. There were no twins explicitly identified in the SMS1932. However, a method for identifying them was devised. It was reckoned that people with the same surname, same date of birth, and in the same school were very likely to be twins. Such individuals were chosen from the SMS1932, and found to have a similar frequency in the population to the twins in the SMS1947. The first question addressed in these twin populations was the 'cognitive cost of being a twin' (Deary, Pattie et al., 2005). There had been equivocal reports in the literature that twins might have lower cognitive test scores than singletons, but no study had examined whole populations. In both the SMS 1932 and the SMS1947 there were very similar results, indicating that, on average, twins were about one third of a standard deviation lower in IQ, which is about 5 points on a traditional IQ-type scale. Further analyses showed that there were no interactions with the family's socio-economic position; twins tended to be lower scorers whatever the parental social class.

A second opportunity provided by these whole populations of twins was examining genetic and environmental contributions to mental ability test scores (Benyamin et al., in press). The initial problem with these analyses was that there were no data on zygosity; it was never indicated whether the twins were monozygotic or dizygotic. To overcome this, a new application was devised of a mixture distribution method of analysis. The reasoning was as follows. The opposite sex twins must be dizygotic. The same sex twins contain a mixture of mono- and dizygotic twins. Given the proportion of opposite sex twins, some estimate can be made of the proportion of dizygotic twins among the same sex twins. Moreover, if the twin-pair differences in Moray House Test scores are plotted, a bell-shaped distribution is obtained. The mixture distribution method tests the hypothesis that this distribution is made up of two different distributions, one for the dizygotic twins and one for the monozygotic twins, with the latter being more peaked and having smaller variance. The application of this novel method gave Moray House Test heritability estimates of 0.73 and 0.67 for the SMS1932 and SMS1947, respectively. The shared environment estimates were 0.16 and 0.26. The non-shared environment estimates were 0.11 and 0.07. These heritability estimates are relatively high for age 11.

Cognitive epidemiology

A novel and fruitful use of the Mental Surveys' data has been in the establishing and developing of a new field, now called 'cognitive epidemiology' (Deary & Der, 2005). Prior to the Mental Surveys' publications there had been urgings that cognitive ability test scores be used in epidemiological investigations, and there had been occasional reports that people with lower IQs tended to die earlier than people with higher IQs (see Deary, in press, for an overview). A study which traced the vital status up to 1997 of almost 80% of the people who took part in the SMS1932 in Aberdeen city found that people with lower IQs tended to die younger (Whalley & Deary, 2001). An exception was deaths in men during active service in World War II; these men tended to have relatively high Moray House Test scores. The effect was not found only at the lowest IQ scores. Indeed, even among the Terman Life Cycle Study of Gifted children, IQ differences are predictive of survival over a 64-year period (Martin & Kubzansky, 2005).

Further lifecourse data were added to some West-of-Scotland participants of the SMS1932 when their Moray House Test scores were linked with data from the Midspan health studies (Hart, Deary et al., 2003). These studies began as health screening studies in the 1970s. 938 people were identified with valid Moray House Test scores from the SMS1932 who also took part in the Midspan studies. After the interview and medical tests in the 1970s (when SMS1932 participants were in their 50s), the participants were followed up using national health databases for deaths and hospital admissions. Thus, the finding that childhood IQ was associated with survival was replicated; people with lower childhood IQ scores tended to die earlier and be less likely to survive in the 25-year period between the Midspan interview in the 1970s and 2002 (Hart, Taylor et al., 2003). This study also showed that the IQ-mortality link was not due to the participants' socio-economic status. Further work on the SMS1932-Midpsan linked data has revealed that lower childhood IQ is linked with cardiovascular disease (Hart et al., 2004), lung and stomach cancers (Hart, Taylor et al., 2003), deaths before but not after 65 (Hart et al., 2005), and hypertension at midlife (Starr et al., 2004).

The link between childhood IQ and mortality is now becoming quite widely replicated (Batty and Deary, 2004; Deary, in press). This has allowed the issue of possible mechanisms for the association to come to the fore. Initially, four non-exclusive mechanisms were proposed (Whalley & Deary, 2001). These hypothesised that IQ might be acting as: (1) an 'archaeological record' of past brain insults; (2) a general indicator of a body's system integrity; (3) a factor leading to higher social class and to safer work and home environments; and (4) a factor leading to better health behaviours. These have been tested only to a limited extent. The first has been explored using the association between birth weight and childhood intelligence (Shenkin et al., 2001,2004); the second by the fact that reaction time accounts for much of the association between IQ and death (Deary & Der, 2005); the third can at best be only a part explanation because there are effects of childhood IQ on survival after socioeconomic status is accounted for (Hart, Taylor et al., 2003); and the fourth has yet fully to be tested, though childhood IQ is associated with giving up smoking (Taylor et al., 2003).

Cohort studies

The largest effort invested in the follow-up studies of the Scottish Mental Surveys has been in the tracing, recalling and retesting of people who took part in SMS1932 and SMS1947. Because of the places where the principal researchers work, these cohorts have centred on Aberdeen and Edinburgh. Recall of the cohorts and their details were described by Deary, Whiteman et al. (2004). Briefly, the recall was facilitated by access to the Community Health Index, the register in Scotland that lists the people who are registered with general practitioners. There have also been calls in the media for volunteers. The cohorts are named after their birth year rather than Mental Survey year. Thus, there are about 230+ and about 500 people in the Aberdeen birth cohorts of 1921 and 1936, respectively. There are about 550 people in the Lothian Birth Cohort of 1921, and the plan is to have about 1000 in the Lothian Birth Cohort of 1936, which is currently being recruited. Many studies have been published using these cohorts to date, and just a selection of these is given below as examples of the sorts of information they provide. The main reason that these cohorts are special is that, in the study of cognitive ageing, they have data on childhood IQ. This is rare, almost unique. If one were to wonder why this is important, consider an analogous situation if we were studying age-related changes in height. If a person were to complain that they were growing shorter we would measure their height and then ask what height they used to be. In cognitive testing, if we were to assess their cognitive faculties in old age it would be useful to know how well they used to score and thus to discover whether their score in old age represented decline, improvement, or no change.

An early task in the study of age-related cognitive change was to find out how strongly childhood IQ correlated with IQ in old age. In fact, Moray House Test IQ scores at age 11 correlated 0.66 (0.73 after correction for attenuation in the range of scores) with scores at age 79 (Lothian Birth Cohort 1921; Deary, Whiteman et al., 2004). This meant that up to about 50% or so of the variance in cognitive ability is stable over the 68 years between 11 and 79. It also meant that there was considerable instability. This instability was a stimulus to our searching for its causes, among various genetic, brain imaging, medical, nutritional, physiological, educational, psychological and social factors. This question is often couched in terms of 'cognitive reserve'; that is, given a person's ability in youth, what factors contribute to their scoring better or worse than expected in later life (Richards and Deary, in press)?

Before describing some of these factors, mention should be made that the cohorts were also able to support the estimating of 'prior' or 'premorbid' IQ in older people by asking them to pronounce the 50 irregular words contained in the National Adult reading Test (NART). The NART score at age 79 correlated 0.65 with the Moray House Test at age 11 (McGurn et al., 2004). Moreover, the almost same correlation and mean score (after correcting for childhood IQ) was found in healthy people and people with mild dementia, validating the supposition that word pronunciation is retained in dementia and provides a valid guide to IQ prior to the illness. It was also shown that the difference between the NART score (an estimate of prior mental ability) and the score on Raven's Progressive Matrices (a measure of current abstract reasoning skill) might be used as an 'instantaneous' measure of the person's lifetime cognitive change, because it correlated highly with their actual cognitive change over that period (assessed using the Moray House Test at age 11 and subtests from the Wechsler Adult Intelligence Scale at age 77) (Deary, Whalley et al., 2004).

Table 1

Genetic studies of age-related cognitive change in follow-up studies of the Scottish Mental Surveys of 1932 and 1947.

Gene	Effect on childhood IQ?	Effect on IQ in old age?	Effect on other cognitive domains in old age?	Reference
APOE	No	Yes	Yes: Memory	Deary et al., 2002, 2004
ACE	No	No	No	Visscher et al., 2004
MTHFR	No	No	No	Visscher et al., 2004
Nicastrin	Yes	Yes	No	Deary et al., 2005
KLOTHO	Yes	Yes[a]	No	Deary et al., 2005
COMT	No	No	Yes: Memory	Harris et al. (2005)
PRNP	No	Yes	No	Kachiwala et al. (2005)
LTF	No	No	No	Kachiwala et al. (2005)
DISC1	No	Yes (women only)	No	Thomson et al. (2005)

[a]This was found in the Lothian Birth Cohort 1921 but not in the Aberdeen Birth Cohort 1936. There was also some evidence that *KLOTHO* variants might be associated with cognitive ageing in women but not in men.

Now, back to some examples of the search for determinants of cognitive change. First, some genetic findings. Table 1 summarises some of the results that have been found to date. There are some general remarks that are worth making. First, the effects tend to be small, with the genetic variants typically accounting for 1-2% of the variance in cognitive test scores.

Second, the samples, even though they have several hundreds of subjects, are quite small for genetic studies, especially when one considers that multiple genes are being tested and that there are multiple cognitive outcomes. Therefore, all of these findings must be taken as suggestive and in need of replication. Third, the cohorts from the Mental Surveys allow a differentiation between genetic effects on life-long cognitive traits and cognitive ageing specifically. Thus, *KLOTHO* and/or Nicastrin variants might be associated with life-long cognitive ability level, whereas *APOE and PRNP* variants might be related to age-related cognitive change. *APOE* and *COMT* might have specific effects on memory in old age. *DISC1* and possible *KLOTHO* might have specific effects on cognitive ageing in women but not men. All of these genes were chosen as candidates because of indications in the literature, but future studies will also address gene systems and whole genome analyses, though these will bring with them additional problems of Type 1 statistical error rates (Deary, Wright et al., 2004).

Brain imaging studies with the Mental Surveys' cohorts have focussed on magnetic resonance imaging (MRI) of white matter, the billions of connections between the brain's nerve cell bodies. A greater number of white matter lesions were shown to be associated with lower cognitive test scores in old age after adjusting for Moray House Test score at age 11 (Deary, Leaper et al., 2003). Hypertension also contributed to lower cognitive test scores, partly via its effect on increasing white matter lesions. A further small study used the newer technique of diffusion tensor MRI to assess the integrity of the brain's white matter. A surprising finding was that, not only was more intact white matter associated with some cognitive functions in old age, it was also associated with cognitive function at age 11 (Shenkin et al., 2003).

In two cohorts, the LBC1921 and the ABC1936, smoking was found to contribute to greater age-related cognitive decline (Deary, Pattie et al, 2003; Whalley et al., 2005). In the latter study this effect persisted after adjustment for education and occupational social class. Education and social class themselves do appear to add some variance to later-life cognitive test scores after adjustment for IQ at age 11, lending some support to the idea of cognitive reserve (Staff et al., 2004). The ABC1936 initially appeared to show that the taking of vitamin supplements was associated with better cognition in old age, but closer inspection showed that this was mostly caused by people with higher IQs at age 11 being more likely to take supplements in old age (Whalley, Fox, Wahle et al., 2004). The level of vitamin B12 was found to be associated with cognitive change between age 11 and 79 in the Lothian Birth Cohort 1921, whereas folate levels were associated with life-long cognitive function, with the association between folate and age 79 Moray House Test scores becoming non-significant once age 11 scores were taken into account (Starr et al. 2005). The importance of having age 11 test scores was found in another study, in which there appeared to be an association between having an earlier menopause and poorer cognitive ageing (Whalley, Fox, Starr et al., 2004). In fact, after adjusting for age 11 Moray House Test score there was no such effect. The conclusion was that women with lower IQs tend to have earlier menopause, rather than early menopause being associated with age-related cognitive decline.

Future studies

Using the data from and following up the Scottish Mental Surveys has contributed some new knowledge to intelligence differences, to the predictive (health-related) validity of intelligence test scores, and to our knowledge of individual differences in cognitive ageing. There are more epidemiology studies to do with the 36-Day and 6-Day samples, which have quite rich social in addition to cognitive data. The cohorts are still being followed up: the ABC1921 and LBC1921 have been seen repeatedly within old age, as have the ABC1936. The LBC1936 is being gathered at present. The DNA and phenotype data stored from the cohorts will provide a valuable resource for testing future candidate genes for cognitive ageing.

References

Batty, G. D., & Deary, I. J. (2004). Early life intelligence and adult health: emerging associations, plausible mechanisms, and public health significance. *British Medical Journal, 329*, 585-586.

Benyamin, B., Wilson, V., Whalley, L. J., Visscher, P. M., & Deary, I. J. (in press). Large, consistent estimates of heritability of cognitive ability in two entire populations of 11-year-old twins from Scottish Mental Surveys of 1932 and 1947. *Behavior Genetics.*

Deary, I. J. (in press). Intelligence, health and death: the new field of cognitive epidemiology. *The Psychologist.*

Deary, I. J., & Der, G. (2005). Reaction time explains IQ's association with death. *Psychological Science*, *16*, 64-69.

Deary, I. J., Hamilton, G., Hayward, C., Whalley, L. J., Powell, J., Starr, J. M., & Lovestone, S. (2005). Nicastrin gene polymorphisms, cognitive ability level and cognitive ageing. *Neuroscience Letters*, *373*, 110-114.

Deary, I. J., Harris, S. E., Fox, H. C., Hayward, C., Wright, A. F., Starr, J. M., & Whalley, L. J. (2005). *KLOTHO* genotype and cognitive ability in childhood and old age in the same individuals. *Neuroscience Letters*, *378*, 22-27.

Deary, I. J., Leaper, S. A., Murray, A. D., Staff, R. T., & Whalley, L. J. (2003). Cerebral White matter abnormalities and lifetime cognitive change: A 67 year follow up of the Scottish Mental Survey 1932. *Psychology and Aging*, *18*, 140-148.

Deary, I. J., Pattie, A., Taylor, M. D., Whiteman, M. C., Starr, J. M., Whalley L. J. (2003). Smoking and cognitive change from age 11 to age 80. *Journal of Neurology, Neurosurgery and Psychiatry*, 74, 1006-1007.

Deary, I. J., Pattie, A., Wilson, V., & Whalley, L. J. (2005). The cognitive cost of being a twin: two whole population studies. *Twin Research and Human Genetics*, *8*, 376-383.

Deary, I. J., Thorpe, G., Wilson, V., Starr, J. M., & Whalley, L. J. (2003). Population sex differences in IQ at age 11: The Scottish Mental Survey 1932. *Intelligence*, *31*, 533-542.

Deary, I. J., Whalley, L. J., & Crawford, J. R. (2004). An 'instantaneous' measure of a lifetime's cognitive change. *Intelligence*, *32*, 113-119.

Deary, I. J., Whalley, L. J., Lemmon, H., Crawford, J. R., & Starr, J. M. (2000). The stability of individual differences in mental ability from childhood to old age: follow-up of the 1932 Scottish Mental Survey. *Intelligence*, *28*, 49-55.

Deary, I. J., Whiteman, M. C., Pattie, A., Starr, J. M., Hayward, C., Wright, A. F., Carothers, A., & Whalley, L. J. (2002). Cognitive change and the APOE e4 allele. *Nature*, *418*, 932.

Deary, I. J., Whiteman, M. C., Pattie, A., Starr, J. M., Hayward, C., Wright, A.F., Visscher, P.M., Tynan, M.C., & Whalley, L. J. (2004). Apolipoprotein E gene variability and cognitive functions at age 79: follow up of the Scottish Mental Survey 1932. *Psychology and Aging*, *19*, 367-371.

Deary, I. J., Whiteman, M. C., Starr, J. M., Whalley, L. J., & Fox, H. C. (2004). The impact of childhood intelligence on later life: following up the Scottish mental Surveys of 1932 and 1947. *Journal of Personality and Social Psychology*, *86*, 130-147.

Deary, I. J., Wright, A. F., Harris, S. E., Whalley, L. J., & Starr, J. M. (2004). Searching for genetic influences on normal cognitive ageing. *Trends in Cognitive Sciences*, *8*, 178-184.

Gottfredson, L., & Deary, I. J. (2004). Intelligence predicts health and longevity: but why? *Current Directions in Psychological Science*, *13*, 1-4.

Gow, A. J., Whiteman, M. C., Pattie, A., Whalley, L. J., Starr, J. M., & Deary, I. J. (2005). Being smart doesn't guarantee happiness: a longitudinal cohort study of the relationship between satisfaction with life in old age and lifetime intellectual function. *British Medical Journal*, *331*, 141-142.

Harris, S. E., Wright, A. F., Hayward, C., Starr, J. M., Whalley, L. J., & Deary, I. J. (2005). The functional COMT polymorphism, Val158Met, is associated with Logical Memory and the personality trait intelligence/imagination in a cohort of healthy 79 year olds. *Neuroscience Letters*, *385*, 1-6.

Hart, C. L., Deary, I. J., Taylor, M. D., MacKinnon, P. L., Davey Smith, G., Whalley, L. J., Wilson, V., Hole, D. J., & Starr, J. M. (2003). The Scottish Mental Survey 1932 linked to the Midspan studies: a prospective investigation of childhood intelligence and future health. *Public Health*, *117*, 187-195.

Hart, C. L., Taylor, M. D., Davey Smith, G., Whalley, L. J., Starr, J. M., Hole, D. J., Wilson, V., & Deary, I. J., (2003). Childhood IQ, social class, deprivation and their relationships with mortality and morbidity risk in later life: Prospective observational study linking the Scottish Mental Survey 1932 and the Midspan studies. *Psychosomatic Medicine*, *65*, 877-883.

Hart, C. L., Taylor, M. D., Davey Smith, G., Whalley, L. J., Starr, J. M., Hole, D. J., Wilson, V., & Deary, I. J. (2004). Childhood IQ and cardiovascular disease in adulthood: prospective observational study linking the Scottish Mental Survey 1932 and the Midspan studies. *Social Science and Medicine*, *59*, 2131-2138.

Hart, C. L., Taylor, M. D., Davey Smith, G., Whalley, L. J., Starr, J. M., Hole, D. J., Wilson, V., & Deary, I. J. (2005). Childhood IQ and all cause mortality before and after age 65: prospective observational study linking the Scottish Mental Survey 1932 and the Midspan studies. *British Journal of Health Psychology*, *10*, 153-165.

Kachiwala, S. J., Harris, S. E., Wright, A. F., Hayward, C., Starr, J. M., Whalley, L. J., & Deary, I. J. (2005). Genetic influences on oxidative stress and their association with normal cognitive ageing. *Neuroscience Letters*, *386*, 116-120.

Martin, L. T., & Kubzansky, L. D. (2005). Childhood cognitive performance and risk of mortality; cohort study of gifted individuals. *American Journal of Epidemiology*, *162*, 1-4.

Maxwell, J. (1969). *Sixteen years on: A follow-up of the 1947 Scottish Mental Survey.* London: University of London Press.

McGurn, B., Starr, J. M., Topfer, J. A., Pattie, A., Whiteman, M. C., Lemmon, H. A., Whalley, L. J., & Deary, I. J. (2004). Pronunciation of irregular words is preserved in dementia, validating premorbid IQ estimation. *Neurology*, 62, 1184-1186.

Richards, M., & Deary, I. J. (in press). A life course approach to cognitive reserve: a model for cognitive aging and development? *Annals of Neurology.*

Scottish Council for Research in Education. (1933). *The intelligence of Scottish children: A National Survey of an age-group.* London: University of London Press.

Scottish Council for Research in Education. (1949). *The trend of Scottish intelligence: A comparison of the 1947 and 1932 Surveys of the intelligence of eleven-year-old pupils.* London, UK: University of London Press.

Shenkin, S. D., Bastin, M. E., MacGillivray, T. J., Deary, I. J., Starr, J. M., & Wardlaw, J. M. (2003). Childhood and current cognitive function in healthy 80-year-olds: A DT-MRI study. *NeuroReport*, *14*, 345-349.

Shenkin, S. D., Starr, J. M., & Deary, I. J. (2004). Birth weight and cognitive ability in childhood: a systematic review. *Psychological Bulletin*, *130*, 989-1013.

Shenkin, S. D., Starr, J. M., Pattie, A., Rush, M. A., Whalley, L. J., & Deary, I. J. (2001). Birth weight and cognitive function at age 11 years: The Scottish Mental Survey 1932. *Archives of Disease in Childhood*, *85*, 189-197.

Staff, R. T., Murray, A. D., Deary, I. J., & Whalley, L. J. (2004). What provides cerebral reserve? *Brain*, *127*, 1191-1199.

Starr, J. M., Pattie, A., Whiteman, M. C., Whalley, L. J., & Deary, I. J. (2005). Vitamin B12, serum folate, and cognitive change from age 11 to age 79. *Journal of Neurology, Neurosurgery and Psychiatry*, *76*, 291-292.

Starr, J. M., Taylor, M. D., Hart, C. L., Davey Smith, G., Whalley, L. J., Hole, D, J., Wilson, V., & Deary, I. J. (2004). Childhood mental ability and blood pressure at midlife: linking the Scottish Mental Survey 1932 and the Midspan studies. *Journal of Hypertension*, *22*, 893-897.

Taylor, M. D., Hart, C. L., Davey Smith, G., Whalley, L. J., Hole, D. J., Wilson, V., & Deary, I. J. (2005). Childhood IQ and marriage by mid-life: the Scottish Mental Survey and the Midspan studies. *Personality and Individual Differences*, 38, 1621-1630.

Taylor, M. D., Hart, C. L., Davey Smith, G., Starr, J. M., Hole, D. J., Whalley, L. J., Wilson, V., & Deary, I. J. (2003). Childhood mental ability and smoking cessation in adulthood: prospective observational study linking the Scottish Mental Survey 1932 and the Midspan studies. *Journal of Epidemiology and Community Health*, *57*, 464-465.

Thomson, P. A., Harris, S. E., Starr, J. M., Whalley, L. J., Porteous, D. J., & Deary, I. J. (2005). Association between genotype at an exonic SNP in *DISC1* and normal cognitive ageing. *Neuroscience Letters*, *389*, 41-45.

Visscher, P. M., Tynan, M., Whiteman, M. C., Pattie, A., White, I., Hayward, C., Wright, A. F., Starr, J. M., Whalley, L. J., & Deary, I. J. (2003). Lack of association between polymorphisms in ACE and MTHFR genes and normal cognitive ageing. *Neuroscience Letters*, *34*, 175-178.

Whalley, L. J., & Deary, I. J. (2001). Longitudinal cohort study of childhood IQ and survival up to age 76. *British Medical Journal*, 322, 1-5.

Whalley, L. J., Fox, H. C., Starr, J. M., & Deary, I. J. (2004). Age at natural menopause and cognition. *Maturitas*, *49*, 148-156.

Whalley, L. J., Fox, H. C., Starr, J. M., & Deary, I. J. (2005). Childhood IQ, smoking and cognitive change from age 11 to 64 years. *Addictive Behaviors*, *30*, 77-88.

Whalley, L. J., Fox, H. C., Wahle, K. W., Starr, J. M., & Deary, I. J. (2004). Cognitive ageing, childhood intelligence and the use of food supplements: possible involvement of n-3 fatty acids. *American Journal of Clinical Nutrition*, *80*, 1650-1657.

Index

A

B

C

D

E

F

G

H

I

K

L

W

Y

ALSO AVAILABLE

HERBERT SPENCER

THE INTELLECTUAL LEGACY

Edited By GRETA JONES AND ROBERT PEEL

Proceedings of the 2003 Conference of the Galton Institute

CONTENTS

ISBN 0950406686

Available post paid from the Institute's General Secretary, price £5.00

A CENTURY OF MENDELISM

Edited by ROBERT PEEL and JOHN TIMSON

Proceedings of the 2000 Conference of the Galton Institute

CONTENTS

ISBN 095040666X

EUGENICS IN AUSTRALIA:

Striving for National Fitness

by

Diana Wyndham

CONTENTS

ISBN 0950406678

HUMAN PEDIGREE STUDIES

Edited by ROBERT PEEL

Proceedings of the 1998 Conference of the Galton Institute

CONTENTS

ISBN 0950406643

Available post paid from the Institute's General Secretary Price £5.00

ESSAYS IN THE HISTORY OF EUGENICS

Edited by ROBERT PEEL

Proceedings of the 1997 Conference of the Galton Institute

CONTENTS

ISBN 0950406635

MARIE STOPES, EUGENICS AND THE ENGLISH BIRTH CONTROL MOVEMENT

Edited by ROBERT PEEL

Proceedings of the 1996 Conference of the Galton Institute

CONTENTS

ISBN 0950406627

POPULATION CRISES AND POPULATION CYCLES

by

CLAIRE RUSSELL AND W M S RUSSELL

CONTENTS

ISBN 0950406651